AGRO-MECHANICAL DIFFUSION IN A BACKWARD REGION

Rakesh Basant
K.K. Subrahmamian

Intermediate Technology Publications, 1990

Practical Action Publishing Ltd
25 Albert Street, Rugby, CV21 2SD, Warwickshire, UK
www.practicalactionpublishing.com

First published 1990\Digitised 2013

ISBN 10: 1 85339 023 2
ISBN 13 Paperback: 9781853390234
ISBN Library Ebook: 9781780441634
Book DOI: https://doi.org/10.3362/9781780441634

Since 1974, Practical Action Publishing has published and disseminated books and information in support of international development work throughout the world. Practical Action Publishing is a trading name of Practical Action Publishing Ltd (Company Reg. No. 01159018), the wholly owned publishing company of Practical Action. Practical Action Publishing trades only in support of its parent charity objectives and any profits are covenanted back to Practical Action (Charity Reg. No. 247257, Group VAT Registration No. 880 9924 76).

The manufacturer's authorised representative in the EU for product safety is Lightning Source France, 1 Av. Johannes Gutenberg, 78310 Maurepas, France.
compliance@lightningsource.fr

Preface

This report is the result of a research project sponsored by the Indian Council of Social Science Research (ICSSR), New Delhi, at the Sardar Patel Institute of Economic and Social Research (SPIESR), Ahmedabad. It is an extensively revised version of the draft report submitted to the ICSSR in early 1984. The comments of the ICSSR referee were received by the authors only in August, 1986; it largely accounts for the delay in revising the report. The data analysed in the report pertain to the year 1981-2; however, the issues raised in the report are still relevant.

In the meanwhile, we have undertaken a similar study in an agriculturally developed region of Gujarat at The Gujarat Institute of Area Planning (GIAP), Ahmedabad. This project has also been funded by the ICSSR. A report on the second project is being prepared and is expected to be ready by May, 1987. To revise an earlier report when a report on a new project on a similar theme is being prepared, is a difficult task. The temptation to report the new findings is strong. We have resisted this tendency and restricted ourselves to the Panchmahals data. Of course, our experience in the developed region has helped us to interpret the results of the earlier survey with a broader perspective. Interestingly, the processes of agro-mechanical technology diffusion are strikingly similar in the backward and the developed regions.

Many people have directly or indirectly contributed to this report. Professor Girja Sharan's direct involvement in the formulation of the project and his later advice as an honorary consultant was very beneficial. At every stage of the project, he has provided useful suggestions and criticism.

Ms Amita Shah had translated the questionnaires from English into Gujarati with a great deal of interest. She never lost sight of the content when deciding about the form of translation. We have also had stimulating discussions with her on our survey findings at various stages of the project.

The Taluka Panchayat officials in Dahod and Kalol helped us in identifying the villages where improved implements are fabricated. The Bhil Sewa Mandal facilitated our field work in Dahod taluka by providing us boarding and lodging facilities at Panchwada Ashram Shala. The love and affection given to us by the Shala Community was overwhelming. Staff members of the Project Planning Cell, Godhra, were equally helpful when we stayed there during our field work in Kalol taluka. Dr PG Pathak, who was in charge of the Godhra Cell at SPIESR, Ahmedabad, was kind enough to provide us with a jeep which made the field work easier. Kaka (Shri Devi Shankar Pandya) never complained while driving on bad roads and at odd hours. Ms Jignyasa Acharya helped us in the arduous task of transferring data from primary schedules to code sheets.

This study would not have been possible but for Shri KL Parmar and Shri Anil Jadav, our investigators, who were as enthusiastic about the project as we were. Their optimism, sense of responsibility and capacity to relate with respondents made the field work an educative and delightful experience.

The main impetus for revising the report has come from Professor Pravin Visaria. He has been a constant source of encouragement and guidance. His comments on the content, organization and detail of the questionnaires were extremely useful. Later he has provided detailed substantive and editorial comments on the draft report, which have improved the revised report in many ways.

We have also had the benefit of detailed comments on an earlier draft from Professor A Vaidyanathan and Shri D Narayana. The comments of the ICSSR referee and our interaction with the colleagues at the GIAP have also helped us a great deal in revising the report.

Shri T Prabhakaran, Ms Sheela Gopal and Shiny Kuriakose have typed earlier drafts of the report. Their systematic, consistent and careful typing has made our obscure, cut and paste manuscripts readable. There are many others, not named, who have assisted with information and comment and provided a great deal of moral support. These include, above all, the respondents who shared their fears, beliefs and knowledge with us.

K.K. Subrahmanian
Centre for Development Studies
Aakulam Road
Ulloor
Trivandrum 695011

Rakesh Basant
The Gujarat Institute of Area
Planning
Sarkhej-Gandhinagar Highway
Gota 382841
Ahmedabad

Contents

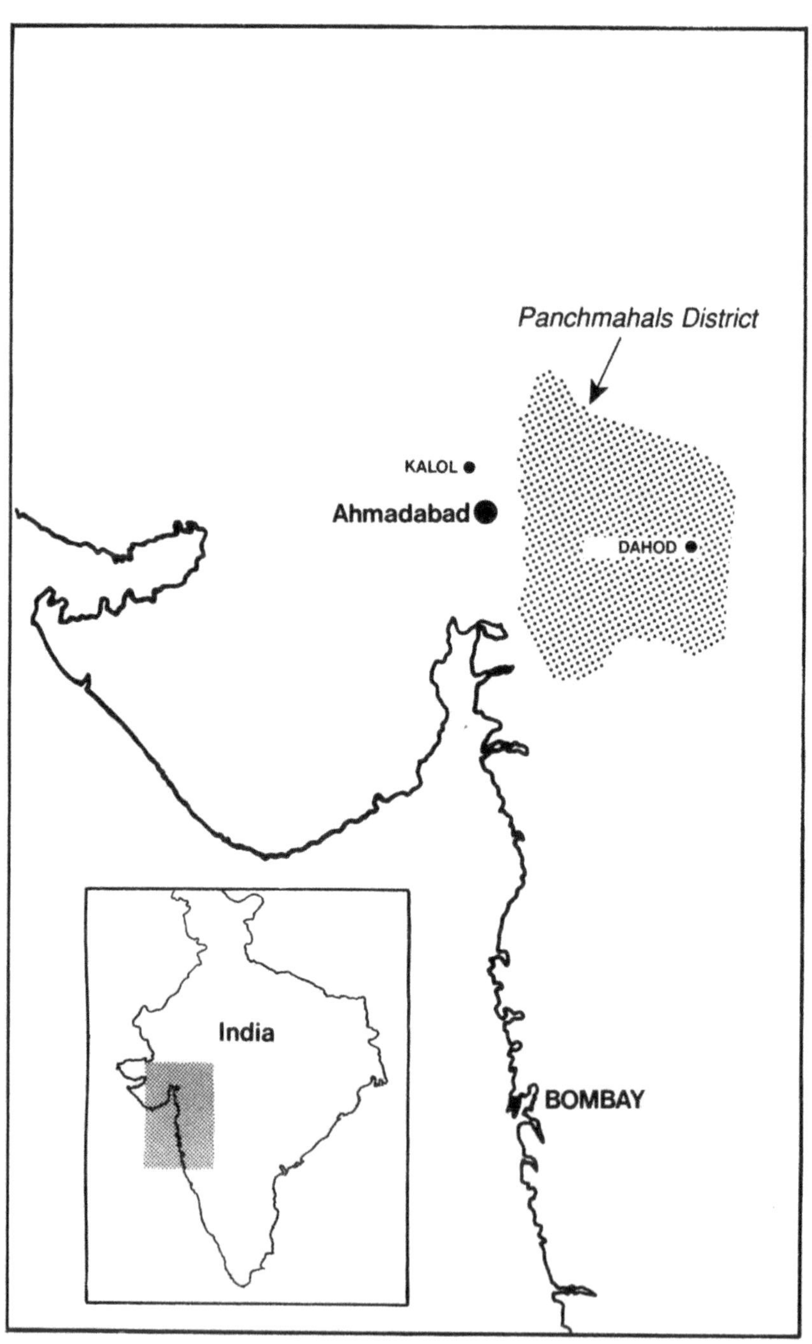

Panchmahals District
KALOL
Ahmadabad
DAHOD
India
BOMBAY

Introduction

Origin and history of artisans in the study region

Historically, agricultural implements were produced within the village by special groups of people who formed specific castes of artisans. Normally village settlements contain all types of artisans: potters, carpenters, blacksmiths, leather workers, etc. This is not always the case, however, as has been brought out in a brief historical review of the artisan families involved in the fabrication of agricultural implements in the study villages.

Sureli: There are two types of artisan families in the village: one set of families is related to two carpenter/blacksmith families who migrated to this village around 150 years ago; others are not carpenters/blacksmiths by caste but have entered this profession recently. Families not belonging to artisan castes have opted for carpentry.

There are four blacksmith families in Sureli, all of them being descendents of the two 'parent' families. Sureli blacksmiths are well known in the region. One of these blacksmiths has stopped working because of old age. Of the six carpenter families in the village, four are related to the parent artisan families and the remaining two are not artisans by caste, i.e. they are not *jat suthars*. Of the four *jat suthars* only two are based in the village and make agricultural implements, the other two do carpentry work (house construction etc.) outside the village. Only the former two were interviewed. One of the two non-*jat suthars* is a harijan and the other belongs to the *Baria* community. These carpenters learnt the trade on their own by watching the *jat suthars*, and were the first in their families to take up carpentry. Both of them mainly made implements for cultivators of their own caste. Of these two, only the harijan carpenter could be interviewed as the other had gone out to work. The ancestors of the harijan carpenter were weavers.

Ghoda: An artisan family from Kalol town was brought by the local cultivators to the village about 75 years ago and given some land. This family was mainly brought in for repair work as most of the farmers had their implements made in Kalol town. Over time the artisans started making new implements for the small and marginal farmers of the region. The parent family split into two households. Both these families are carpenters/ blacksmiths, partly because of the fact that they basically do repair work. Unlike Sureli carpenters, these artisans do not go out to work on house-sites.

Gangardi: Of the four blacksmiths in the village, two have long associations with the village while the other two have started working in the village recently. One of the old blacksmith families came to the village more than 250 years ago. Since this family was the only blacksmith family in the region when the British took over (1860), they were given some land which was registered in 1955-6. Only one person from this family now works as a blacksmith. Due to old age, he does repair work only.

The other blacksmith family came about 50 years ago from Limkheda taluka of the same district, Panchmahals. The son of the artisan who migrated is the only person who works as a blacksmith in the family now, others having shifted to retail trade. He too is old and does mainly repair work. The remaining two blacksmiths do not stay in Gangardi: they commute everyday from nearby villages (two to three kilometres away) and work in rented sheds. They have been working in Gangardi for the last three to four years. One of them also knows carpentry as he worked with a furniture-making group earlier.

There is no *jat suthar* in Gangardi. Cultivators of all castes, except *brahmins*, *nais* (barbers) and *veparies* (businessmen) make their own ploughs and yokes, the two main wooden implements in use. Some tribal households have started doing carpentry work. We interviewed only one artisan who makes wooden agricultural implements for others. There is one more carpenter who goes to Dahod everyday for work and does mainly house work. Another *bhil* carpenter had died recently and his son has not taken up his vocation. The carpenter interviewed is basically a farmer. Since his sons looked after cultivation, and since crop income was not adequate for his family, he decided to make agricultural implements for others. Instead of going for wage labour (like his sons) he supplemented the crop income by carpentry. About ten years ago, he started working on house construction sites as well.

In Gangardi, there is another set of harijan households (four) making agricultural implements of bamboo, such as seeding funnels (*orni*). These families live in houses located away from the main village and are related to each other. They reported that they are as old as the village. Earlier these families used to get foodgrains from most of the households of the village, as a custom. Now they supplement their incomes by agricultural wage labour. One of these households was interviewed.

Dadur: In Dadur, there is neither blacksmith nor *jat suthar*. There is only one cultivator who does carpentry work for payment. He is the first in his family to do carpentry work; his forefathers were cultivators. He started by fabricating agricultural implements but now does construction work as well.

Apart from this carpenter, there are four or five households, situated away from the village, involved in leather work. These families belong to a harijan caste known as *Makwana*. These households have their origins in the same family which became divided over time. The main agricultural implement they make is the *moht*, a leather bag used for irrigation. These households also supplement their incomes by wage labour.

Apart from the artisans based in the selected villages, three others producing implements on a larger scale with modern fabrication facilities were also interviewed. One of these artisans, a blacksmith, was based in the village Garbada in Dahod *Taluka* and the other two in Kalol town.

Garbada: Cultivation was the main occupation of the forefathers of the Garbada artisan. They did some blacksmithy only as a subsidiary occupation and made a very limited set of implements, mainly hand tools. The present artisan started with repairing some sophisticated implements like the iron-blade harrow and Baroda hoe. About five years ago, he took a loan and bought some modern tools, including a welder and grinder. He now makes improved implements which qualify for subsidy.

Kalol: The forefathers of the two families here did carpentry as well as blacksmithy. Over time, one family specialized in carpentry and the other in blacksmithy.

The family members who expanded the business from that of artisanal to mechanized production had experience of working in factories. One of the forefathers of the blacksmithing family received a certificate of merit in the Industrial and Agricultural Exhibition in Bombay in 1904. Another helped to establish a workshop in a Panchmahals steel factory. This family was the first to own a lathe in Kalol. They started by making steel furniture such as safes, almirahs, etc., and have diversified to iron agricultural implements only recently. They now qualify for subsidy schemes.

The carpenter proprietor also worked for many years as a turner and instructor in railway workshops and private sector factories before he expanded his own trade by purchasing some mechanized fabrication facilities. Both the families now have ample equipment for outside jobs on a contract basis, such as cutting wood, welding and forging.

Thus, one finds that, while the tradition of specialized blacksmithy is a long one in all the study villages, only Kalol villages have had access to specialized carpenters. The carpenters in the Dahod villages are not carpenters by caste, but have acquired these skills over time.

The agricultural implements of the study region

Forty-eight types of agricultural implements were enumerated in our primary survey. A few of these were identical to each other except for the material used in making them. Some implements were multi-purpose in nature while others were used for specific operations.

The implements used in a particular region are directly affected by prevailing agricultural practices. The agricultural practices and consequently the implements used reflect the level of agricultural development in a region. Since agricultural implements contribute to the various phases in the process of agricultural production, in what follows we briefly review the role played by the various agricultural implements of the region in the process of agricultural production. In the discussion, an attempt is also made to bring out inter-village differences in agricultural techniques.

Implements for land preparation: Preparatory tillage serves several functions. Apart from preparing a proper seed-bed it is meant to increase the capacity of the soil to absorb and retain moisture, to prepare the land for efficient use of irrigation water, to uproot weeds, as well as to control their subsequent emergence. It consists of a number of distinct operation such as ploughing, harrowing, levelling and bunding, in varying combinations and intervals depending on the agro-climatic and soil conditions and also the nature of the crop. In what follows, we describe agricultural implements in the most common sequence of agricultural operations.

Immediately after the first rains, cultivators in Kalol villages use blade harrows (*Karab*) for primary tillage. This loosens and breaks the soil, and destroys weeds. The cultivators of the Kalol villages argue that the use of ploughs for primary tillage leaves the land 'cloddy' and reduces its water-retaining capacity.[1] Normally, a plough is used after softening the land with the blade harrow but in some cases the plough alone is used. The latter type of practice is prevalent in Dahod villages where blade harrows are not used. Tractor-driven cultivators are also used by some cultivators in all the selected villages except Sureli. In cases where both (tractor-drawn cultivators and bullock operated ploughs) are being used, the cultivator is used first so that use of the plough may be easier.

For paddy cultivation, land is prepared for transplantation by the use of bullock-drawn puddlers (*ghaniyun*). This implement is used only in irrigated fields which can be flooded with water or in low-lying (*kyari*) land where rainwater accumulates. In Dahod villages, the puddler is normally used in *kyari* land. The lack of rains during the last two to three years has reduced use of the *ghaniyun* in the selected villages of Dahod.

Land-levelling implements: After tillage, the next step in seed-bed preparation is land levelling. Various types of clod crushers, levellers and smoothers are used for this purpose. Planks (*samar*) of various sizes are used in all the selected villages to level the land and break the clods if there are any. The planks also press the land which helps in water retention. If the land is very undulating, a scoop (*petari*) and scraper (*kain*) are used to level the land. This type of levelling is required to stop rain and irrigation water from flowing out of the fields. Scoops and scrapers are found only in Dahod villages, where the land is undulating; the land in Kalol villages is, however, uniform and plain and these implements are not needed there.

Seeding implements: Broadcasting, transplanting and row-sowing are the three most prevalent ways of sowing seeds in the study villages. Dibbling of seeds was also practised by cultivators in the Kalol villages, especially for HYV maize.

As already mentioned, transplanting is normally done for irrigated paddy or paddy grown in *kyari* lands. Otherwise, paddy is broadcast, as are some coarse grains, particularly when the rains are insufficient. Some oilseeds are also broadcast; they are usually thrown in those fields where some other crop has already been sown. Broadcasting is more prevalent in Dahod villages as compared to Kalol villages.

No implement is needed for broadcasting the seeds. Land preparation may be done with the help of a puddler for paddy transplantation. Three sets of implements are used for row sowing:

○ Two-tine seed-drill and a seeding funnel (*padko* and *nalichala*)[2]
○ Three-tine seed-drill and a seeding funnel (*tarphan* and *nalichala*), and
○ Plough with a single-hole seeding funnel (*orni*).

In Kalol villages all three sets are used (the first two being more common) while only the third set is used in Dahod villages. The two- and three-tine seed-drills have a fixed distance between two tines; in the case of a plough, the distance between rows can be changed at will, although it is somewhat difficult to maintain uniformity in the distance between rows. The distance between crop rows was found to be smaller in Dahod villages than in Kalol villages.

Sometimes planks (*samar*) are used after seeding to press and cover the seeds properly.

Implements for manuring and applying chemical fertilizers: Manuring is normally done at the time of land preparation. Bullock-drawn carts or tractor-driven trolleys are used to bring it to the field and rakes (*panjethi*) and hand hoes/shovels (*pavda*) are used to spread it. Chemical fertilizer is applied along with seeds or after sowing. Seed-drills are used by some for di-ammonium phosphate, particularly when it is mixed with seeds. Single-hole funnels are used by some cultivators to apply urea along the crop rows and alongside plants when they are small. Use of seed-drills/funnels entails a saving of chemical fertilizers. This method is mainly adopted in Kalol villages, especially Ghoda. Chemical fertilizer is normally broadcast in Dahod villages.

Irrigation implements: Leather or iron *mohts* (*chadas*) are used to lift water for irrigation in Dahod villages. Oil engines are gradually becoming popular.

A man, a pair of bullocks and *tareleyun* (a kind of yoke) are needed to use the *moht* to lift water. Electric motors, and to a limited extent oil engines, are used in Kalol villages. Bunding for irrigation is done by bullock-drawn bund-formers (*bambphario*) by some Ghoda cultivators. In other villages, rakes and hand hoes are used for this purpose.

Implements for weeding and intercultivation: Hand weeding is normally done with sickles in the selected villages. Intercultivation is done by bullock-drawn blade hoes (*karbadi*) in Kalol villages.[3] Sickles are used here mainly to weed within the same row. The size of the blade of the blade-hoe is generally determined by the distance between the tines of the seed drills because that determines the distance between plant rows. A small blade-hoe, however, can be run twice between the same two rows for intercultivation if the distance between the two tines of the seed-drill is very large.

Since the distance between crop rows is generally small in Dahod villages, blade-hoes are very rarely used. Most of the weeding is done manually by sickles. The practice of intercultivation is adopted by a limited number of farmers and is performed with a plough.

Plant protection devices: Dusters and sprayers are used for dry and wet pesticides. Some cultivators in Dahod villages broadcast pesticides with their bare hands.

Harvesting implements: Grain crops are harvested by means of sickles in all the four villages. The sickles, however, are different. While the Kalol sickles are serrated, light and small, the Dahod sickles are non-serrated, heavy and large. This is partly because the sickle in Dahod villages is also used for cutting fodder and branches of wood and for other purposes which are performed by bill hooks and other implements in Kalol villages.

Cotton is picked by hand in all the villages but its stalks are pulled out by some cultivators in Ghoda with a cotton plant puller (*cheepeo*). Most of the cultivators, however, use digging hoes (*kodali*) for this purpose. The use of the cotton stalk pullers has declined because of the decline in the area under cotton; some cultivators in our sample who owned this implement did not grow cotton.

In Sureli (a Kalol village), a *karban* (a bigger, heavier and slightly different version of blade-harrow) is used to harvest groundnuts. Blade harrows are used for this purpose in Ghoda, the other Kalol village.[4] The cultivators in Dahod villages use ploughs for this operation also.

Threshing and winnowing implements: Threshing is normally done with the help of bullocks, who are made to walk on the harvested crop. Some cultivators in Ghoda have replaced bullocks with tractors for this operation. Rakes are used to separate grass and hay from the grain while bullocks are walking on the harvested crop. In Dahod villages, a special rake (a traditional implement called a *kadial*) is also used for this purpose.

For crops like maize and groundnut, threshing is normally done by hand in Sureli and Dahod villages. Sometimes, sticks are used to beat maize cobs to separate grain. Threshers and tractors are also used for threshing; these services are mostly hired. For winnowing, some cultivators in Ghoda use winnowing fans while in other villages thick sheets of cloth are used to generate wind for this purpose.

Transport facilities: Bullock-drawn carts and tractor-driven trailers or trolleys are used to transport grain from the field to house or market, as the case may be.

Miscellaneous hand tools: A few cultivators in Ghoda use a hand chaff-cutter (*sudo*) to cut fodder, but most cultivators in Kalol villages use axes for this purpose. In Dahod villages, sickles are used to cut fodder. The remaining implements are multi-purpose in nature. The digging hoe (*kodali*), pickaxe (*trikam*) and crowbar (*narash/rambho*) are basically digging implements. The digging hoe is utilized for general digging, pickaxes are used for harder surfaces, and crowbars are used mainly to dig holes for fencing. Axes, apart from general wood cutting, have another interesting use: small pieces of wood are cut and used to alter the angle of some bullock drawn implements e.g., the angle of the plough share or the blades of blade harrows and hoes. This method is also utilized to alter the distance between the tines of a seed-drill.

Notes

1. It is noteworthy that the blade-harrow is a unique implement in India, not found anywhere else in the world. Within India also it is used in the semi-arid tracts of Gujarat, Madhya Pradesh, Maharashtra, Andhra Pradesh and Tamil Nadu (India, NCA 1976, pp.397-98).
2. *Nalichala* consists of a wooden/iron bowl with holes and wooden/iron tubes which are fitted in these holes. Plastic and tin versions of this accessory are also available.
3. Blade hoes are also implements found only in India. These are very popular in Gujarat (India, NCA, 1976, p.404).
4. It should be noted that blade harrows in Ghoda are bigger and heavier compared to the blade harrows in Sureli.

Chapter 1
The Framework

An understanding of technology development and diffusion can contribute to the identification of some of the basic forces operating in a society. The interest of economists in this development of technology lies in its importance as an indicator of economic growth. Historically, technological progress has varied over time and space: the capacity of different societies to generate, adapt and disseminate appropriate technical innovations has shown considerable variation. The reasons for these differences reflect the socio-economic structure of these societies.

Most people in developing countries live in rural areas and work in agriculture. In India, after three decades of planning for socio-economic development, 76 per cent of the population lives in villages; agriculture accounts for about 42 per cent of the national income and provides a livelihood for more than 67 per cent of the population (India, Planning Commission, 1981; India, Registrar General, 1982). Rural development and the improvement of the low level of agricultural productivity have, therefore, been priority concerns of Indian planners since 1950. Development and diffusion of agricultural technology has been an important source of agricultural growth. Many studies have analysed the diffusion of agricultural technology and its impact on output, yield, employment and income distribution. There are, however, very few studies on the diffusion of agro-mechanical technology (Feder *et al.*, 1982). Moreover, the available studies mostly concentrate on relatively expensive equipment such as tractors, threshers, oil engines and electric motors. Much less has been said about simpler and less expensive equipment whose contribution to agriculture may not be as visible but is still important, both in terms of lessening the drudgery of work and higher labour efficiency. Besides, the viability of mechanization can be questioned on many grounds in a country such as India where the majority of land holdings are small, 67 per cent of rural households owning less than one hectare of land (NSSO, 1986).

The agro-machinery used on various farms differs across regions in accordance with the nature of cultivation. Broadly speaking, there are two sources of farm-power: tractor-based agro-machinery and bullock- or man-based. In a given context, larger holdings use different sets of agro-machinery and, therefore, different types of farm power, than smaller ones. The ownership pattern of tractors and tractor-based equipment is heavily in favour of the large holdings.[1] The picture is similar in the case of electric motors and diesel engines (Dasgupta, 1977, Chapter 2). The growth of tractor-hiring facilities enables some of the smaller farmers to use tractors on rent, yet small farmers (with less than two hectares of land) are unable to make use of the facilities on any significant scale (Sharan and Krishna, 1976; Patel, 1980). Much of the same imbalance will naturally be seen in agro-machinery and equipment that is linked to the tractor as a power unit. Also, the system of hiring farm machinery, such as tractors and harvesters, has given rise to a new type of patron-client dependency between poor and rich farmers (Dasgupta, 1977, p.379).

There have been serious R&D efforts to develop small tractors. However, some technical problems make it difficult to develop a low horsepower tractor that would be sturdy and efficient and also affordable to small and marginal farmers. Besides, tractors are in fact not suitable in all conditions: bullocks perform better than tractors on clay soils, in ploughing corners and in isolated patches of highland. Moreover, tractors do not usually displace bullock labour. It has been found that tractor farms continue to keep at least one pair of bullocks (Dasgupta, 1977, pp.108-9; Raj, 1974, p.123). Thus, tractor-hire schemes notwithstanding, tractor-based agro-machinery remains limited to upper segments of the peasantry. The marginal and small farmers, who constitute a large proportion of the total cultivator population, rely mainly on bullock/man-operated implements. In fact, it has been shown that in the near future, available farm power will be dominated by manual and draft power; the tractor population, on the other hand, is small and only a modest increase is expected in the coming years (India, National Commission on Agriculture (NCA), 1976, Part X, Chapter 50).[2] Furthermore, the impact of machines such as tractors, threshers, electric motors and diesel engines is similar to that of seed fertilizer technology. If these services are available only to a limited set of cultivators, the biases of seed fertilizer technology become a policy concern. The other major concern has been that mechanization may reduce employment. There is, however, no consensus regarding the impact of the use of above-mentioned machines on yields, output and employment (Binswanger, 1978; Basant, 1985).[3]

Large scale co-operation, consolidation of small holdings, socialization and reorganization of land into holdings of optimal size can overcome structural constraints and pave the way for more rapid mechanization. It has been argued, however, that small farms which constitute a large chunk of the peasantry will continue to remain a very important part of the agricultural scene in India (Vyas, 1976; Joshi 1979). It is imperative, therefore, to think of ways in which technological upgrading of agro-machinery on bullock-operated farms can be accomplished. It is noteworthy in this context that some of the improved hand-operated and animal-drawn implements can be five to ten times more efficient in energy conversion as compared to primitive tools and implements commonly used by farmers (India, NCA, 1976, p.393). It has also been pointed out that improved implements can greatly accelerate bullock ploughing (Marsden, 1973, p.7). It is in this context that the present study attempts to analyse the processes of agro-mechanical technology diffusion in a backward region (Panchmahal district of Gujarat State).

Focus of the study

Development and diffusion of agro-mechanical technology characterized by improved hand- and bullock-operated implements, therefore, is an important area of research. Few studies have, however, focused on this theme (for a detailed review see Feder *et al.*, 1982). The present study is an attempt in the direction of understanding the constraints on the diffusion of agro-mechanical technology. The relevance of the study in the present context is brought out by an awareness that historically successful agricultural growth in different countries has capitalized on favourable production factors, e.g. land and mechanization in the US; labour and land improvements and biological technology in Japan. European countries also stressed biological technology before emphasis shifted to mechanical technology (Binswanger, 1984). Further, to rely mainly on mechanization for output growth may be misplaced since maximizing yields at the margin is not the same thing as maximizing output in a country as a whole. The two objectives can diverge widely if the former means concentrating scarce capital and skill resources on land with the greatest natural endowments, and on those holdings which can afford mechanization, while neglecting or discouraging more modest improvements which could be reproduced much more widely (Marsden, 1974, pp.5-6).

In a broad sense, technology is more than just machinery and equipment or knowledge of methods of production; it is an integral part of the whole system of economic, institutional, social and even political arrangements that characterize the way a society functions. Looked at in this fashion, the role that technology can play will depend upon the socio-economic and political structure of the society and its transformation over time. In the context of agricultural technology, therefore, it is important to analyse not only its relation to output and inputs but also its inter-relationships with the institutional infrastructure of the region, the extent of commercialization of farming, the non-agricultural activities in the region and the values, attitudes, customs and behaviour patterns of rural society. Economic, social and other institutional adjustments thus become an integral part of technological change (Wang, 1981, p.7). An exploration of these relationships with specific reference to agro-mechanical technology is the focus of the present study.

As mentioned earlier, the agro-machinery situation in India is characterized by two distinct technological worlds: a small segment consisting of large farmers increasingly employing tractor-powered equipment; and a large segment of small peasants using only hand- and bullock-operated equipment. Research efforts seek to improve and modernize this machinery but the pace of adoption of improved implements is extremely slow (India, NCA, 1976). Most tools are still designed, fabricated and supplied by local artisans and markets, particularly in backward areas.

Diffusion of technology is not, however, merely the pace of adoption or spread of a new technology, as normally referred to in economics. To the extent that technologies are difficult to imitate and require learning-by-doing and R&D, the technological development of a region should be viewed as its capacity to generate, transfer, adopt, adapt and incorporate into the economic system new techniques which displace or modify the older techniques, and/or to facilitate these processes in a smooth manner. Technology diffusion refers to this whole process. Defined in this sense, diffusion is a continuous process and is central to socio-economic development. The motivations and mechanics of diffusion are subjects that have received considerable attention. Analysis, however, has focused on changes in the structure of demand and on changing relative cost factors as the main inducement. Economists have viewed the process of diffusion as a response to profit expectation, as shaped by market size and structure. (For a detailed review of studies on agricultural technology diffusion, see Feder, *et al.*, 1982).

While the demand side is important, diffusion of technology also depends upon the pace at which supply-side constraints are overcome. Relevant supply-side factors include the state of scientific knowledge, prevailing technology, changes in the number of skilled workers engaged in production and the upgrading of their skills, the organizational structure of technology generation, links between different agents involved in technology generation and the economic system. These factors determine the endogenous R&D and engineering efforts which may be called local technological capacity. Without an understanding of these supply-side forces, it is difficult to explain technology diffusion in a developing country, particularly in a backward region.

Methodology

This study is based on an intensive survey of four villages belonging to two different *talukas* of a backward district in Gujarat. It is essentially a micro-level study of the process of technology diffusion in agriculture. The possibilities of drawing generalized inferences are obviously limited. In so far as the selected villages represent different levels of development, one may be able to generate testable hypotheses regarding the processes of technology diffusion in transitional agrarian economies.

The district selected for study, Panchmahals, is regarded as one of the most backward districts of Gujarat in terms of agricultural development. Yield levels in the district have been low with limited growth of agricultural output. In the 1960s the compound annual rate of growth of agricultural output was less than 1.5 per cent (Bhalla and Alagh, 1979, Appendix 5), without any significant improvement in recent years (Sharma, 1983).

Villages from two *talukas*, Kalol and Dahod, were chosen for the study. Indices of agricultural productivity show Kalol to be one of the better *talukas* and Dahod as one of the not so good ones (Government of Gujarat, 1979). While Dahod is a dry and agriculturally backward region, Kalol is relatively better irrigated and, agriculturally, better endowed. A comparative study of these contrasting *talukas* within the same district was expected to yield some useful insights into the specific processes of technology diffusion.

Selection of sample villages

Officially published sources did not yield any information about the number of artisans making agricultural implements in the various *talukas* and villages of Panchmahals district. *Taluka* Panchayat officials of Dahod and Kalol *talukas* were contacted to identify those artisans who make improved implements on a somewhat large scale and supply to the neighbouring villages. Jesawada and Garbada were identified as the two such centres in Dahod *taluka*, and Kalol town and Derol in Kalol *taluka*. It was then decided to concentrate efforts on the regions covered by the Kalol town artisans in Kalol *taluka* and Garbada artisans in Dahod *taluka*.[4] Although Kalol and Garbada 'artisans' mainly produce agricultural implements only on order, their production process is highly mechanized and they also sub-contract for other manufacturers.

The artisans making improved implements in Garbada and Kalol town were contacted and were asked about the villages they cater to in the area. Two villages in each *taluka*, one backward and the other developed, were selected from this list. In the absence of any detailed village level information on the levels of agricultural development, the selection was based on the extent of irrigation using cultivated area per electric pump/oil engine as a proxy (Table 1).[5]

Selection of sample households

Cultivators who use agricultural implements, artisans who fabricate them and registered agents of outside fabricators in the market were identified as the three main agents directly involved in the process of generation and diffusion of agro-mechanical technology. Three questionnaires were created, one each for cultivators, artisans and agents. As an example, the questionnaire for artisans is included as Appendix 1. All households in the four selected villages were listed. For selecting a sample of cultivators, all cultivator households were first divided into four strata of operational land holdings: marginal (less than one hectare); small (1-3 hectares); medium (3-7.5 hectares) and large (more than 7.5 hectares) for the four study villages. A stratified random sample of 20 cultivators was then chosen from each village. The number of sample households constituted more than 15 per cent of the total cultivator

Table 1: Cultivated area per electric pump and oil engine in 1977 in the villages serviced by the contacted artisans of Kalol and Dahod *Talukas*

Kalol *Takuka*		Dahod *Taluka*	
Name of village	Cultivated area per electric pump and oil-engine (acres)	Name of village	Cultivated area per electric pump and oil-engine (acres)
1. Adadra	114.8	1. Abholod	—
2. Alindra	—	2. Boriala	—
3. Bakrol	—	3. Chandle	183.2
4. Bedhia	177.3	4. Dadur*	—
5. Baru	175.8	5. Derdha	—
6. Dekol	159.7	6. Gangardi*	111.0
7. Derol	18.4	7. Garbada	626.3
8. Hgoda*	16.1	8. Jambua	558.0
9. Khatol	168.5	9. Jesawada	172.8
10. Malav	28.4	10. Kharedi	573.0
11. Medapur	152.6	11. Nandwa	—
12. Rabad	20.6	12. Nelsur	—
13. Ratanpura	65.5	13. Zari-Bujarg	3441.0
14. Sureli*	-		
15. Tarwada	45.1		
16. Vejalpur	24.2		

* Villages selected for detailed survey.

Sources: 1.   *Census of India* (1971), Series 5, Gujarat District Census Handbook, Panchmahals District, Part X A & B, Town and Village Directory, Village and Townwise Primary Census Abstract.

          2.   *Livestock Census* 1977.

households in the village, except in Sureli, where it constituted only 6.5 per cent.

All the blacksmiths and carpenters in the selected villages were surveyed, along with the artisans making improved implements in Kalol town and Garbada. All the agents operating in the selected region were also interviewed.

Notes

1. A survey of tractor ownership in the states of Punjab, Haryana, Uttar Pradesh, Andhra Pradesh, Tamil Nadu, Maharashtra and Gujarat has revealed that only 1.3 per cent of farmers of less

than 2 hectares and only 4.2 per cent of those with 2 to 4 hectares own tractors. The proportion of tractor-using households is, however, much higher; 64 and 79 per cent respectively for these groups of holdings (NCAER, 1981). This may partly be because the NCAER data represent only eight, relatively developed, states.

2. For a brief discussion of bottlenecks in tractor production, see Dasgupta (1977), pp.109-10.

3. There are indeed analytical problems in separating out the impact of machines such as tractors and threshers from that of electric motors, HYV seeds etc. On balance, however, the available evidence suggests that all types of irrigation (including mechanized pumping of water) and bio-chemical inputs contribute to increases in yields and in labour input in total crop production. The use of threshers, tractors and harvesters displaces labour. The negative impact of such mechanization on labour use comes out more clearly at the level of individual crops and the overall intensity of this decline is determined by the operations mechanized (See, for a detailed review of studies on these issues, Basant, 1985).

4. The term 'artisans', as used here, does not fully conform to its traditional connotation of special groups of people which have historically evolved as special castes and as part of special division of labour.

5. While the figures for electric pumps related to 1977, the cultivated area figures refer to 1971. It was assumed that the cultivated area in 1977 was more or less the same as in 1971.

Chapter 2
Socio-economic Profile of the Study Villages

Panchmahals district is located on the eastern border of Gujarat State. Five of the eleven *talukas* and about 42 per cent of the district population are tribal. At the turn of the century, a large part of the district area was under forest or cultivable waste. In 1880-1, only 27 per cent of the total area was under cultivation, about 36 per cent under forest and 33 per cent cultivable waste. A large proportion of the uncultivated area was located in tribal *talukas*. Since then, however, the area under cultivation has steadily increased and, in 1974-75, the net sown area accounted for about 54 per cent of the total reported area of the district. Thus, agriculture in Panchmahals district, particularly in the tribal *talukas*, does not have a very long history. Of the two *talukas* chosen for this study Dahod is a tribal *taluka* while Kalol is non-tribal. According to the 1981 Census, the proportion of tribal population was 73 per cent in Dahod and only 8 per cent in Kalol (Government of Gujarat, *Panchmahals Gazetteer*, 1972, Statement IV.2, p.267; India, Registrar General, 1981, Paper 2; District Statistical Abstract, Panchmahals, 1974-75).

The major part of the district is 'hilly' and 'undulating' and Dahod *taluka* belongs to this part of the district. Kalol *taluka*, on the other hand, is flat (Government of Gujarat, *Panchmahals Gazetteer*, 1972). Both *talukas* have a dry climate, with the south-west monsoon providing rain during June-September. In terms of average annual rainfall, however, Kalol *taluka* is slightly better than Dahod.[1] Panchmahals is a drought-prone district. Historically the eastern portion of the district of which Dahod is a part has been more vulnerable to scarcity conditions than the western portion where Kalol is located.[2] In recent droughts also Dahod has been one of the worst attested *talukas* of the district. Kalol on the other hand is relatively less prone to droughts and famines.[3] The villages selected from these two *talukas* provide a contrasting picture in terms of their socio-economic profile.

Infrastructural facilities

The infrastructural profile of the study villages is summarized in Table 2. The

Table 2: Infrastructural facilities in the surveyed villages, 1981-2

	Items	Ghoda	Sur#li[a]	Gangardi	Dadur
			Villages		
1.	Distance from nearest town (Km)	1 (Kalol)	13 (Godhra)	10 (Dahod)	12 (Dahod)
2.	Distance from the nearest market (Km)	1 (Kalol)	3 (Vejalpur)	0 (Gangardi)	2 (Gangardi)
3.	Communication	WY	WY	WY	FS
4.	Electricity	Yes	Yes	Yes	No

Bus Facility: WY - Whole Year,   FS - Fair Season
a. One part of Sureli can only be reached by a fair weather track suitable for 4-wheel drive vehicles.

village Ghoda in Kalol *taluka* is about one kilometre off the Godhra-Baroda highway. The road connecting the village to the highway was being converted into an all-weather road while our survey was being conducted. Ghoda is easily accessible by road and a number of local buses plying the Godhra-Kalol route stop at the village. Besides, Kalol town, which is well served by roads is only one kilometre away. Sureli, the other village selected in Kalol *taluka*, though not on the highway, has easy access to transport facilities.

The village Gangardi, selected from Dahod *taluka*, has a bus service. An all-weather road connects Dahod, the *taluka* headquarters, and Gangardi. For Dadur, which is about 2.5 kms from Gangardi, there is only a fair-weather road. Very few buses go to Dadur and even this infrequent bus service is suspended during the monsoon months.

In terms of access to the market for agricultural inputs and outputs, Ghoda is well situated because of its proximity to the *taluka* headquarters, Kalol. For Sureli, the other village selected from Kalol *taluka*, the closest market is in Vejalpur village which is 3-4 km away. Gangardi, in Dahod, is a big grain centre with a large co-operative (Gangardi Vibhag Krushi Sewa Sahkari Mandli Ltd) handling the bulk of grain-trading activities in the region. Although the Gangardi-based co-operative has recently opened a retail outlet at Dadur, the village is dependent on Gangardi for all its market needs, including cloth and clay vessels. Processing of grain also has to be done in Gangardi because there is not even a flour mill in Dadur. Sureli in Kalol has its own flour mills, potters, tailors and retail traders, while Ghoda gets all these items from Kalol town.

Nature of agriculture

Inter-village differentials in agricultural productivity, irrigation, cropping pattern and intensity, use of bio-chemical inputs and rates of commercialization are shown in Table 3. Agricultural productivity is highest in Ghoda, followed by Sureli, Gangardi and Dadur, in that order. Differences in cropping intensity across villages are, however, not that marked. In fact, the Dahod villages (Gangardi and Dadur) show a somewhat higher cropping intensity. The high agricultural productivity in Kalol villages, particularly Ghoda, is mainly due to differences in the use of irrigation, fertilizer, HYV seeds and in cropping pattern. The proportion of irrigated land is much higher in Kalol villages, particularly in Ghoda, compared with Dahod villages. Wells are the major source of irrigation but, as shown later, electric motors and oil engines are used more widely in Kalol villages (especially Ghoda), while Dahod villages mainly use *moht* (leather water bags) and to some extent oil engines. A similar picture emerges as far as the consumption of chemical fertilizers and pesticides is concerned. Use of traditional manure, however, is still widespread.

Foodgrains are the predominant crop in all four villages. The paradoxical situation of developed Kalol villages, especially Ghoda, having a higher proportion of cultivated area under coarse cereals than the less-developed Dahod villages can be easily explained. The category of coarse cereals includes summer *bajri* (pearl millet) which is actually a cash crop. HYV *bajri* seedlings are produced in this crop. All the female component of the crop, which forms more than 80 per cent of the total output, is sold in the market. It is a high value crop, its price being three times higher than that of monsoon *bajri*. Higher agricultural productivity in Kalol villages is also explained by a higher proportion of other high-value cash crops such as groundnut and cotton. Another important difference is that while in Kalol villages the medium/small holdings have a more diversified cropping pattern as compared to other holdings, in Dahod the same is true of large holdings. This indicates that small and medium farmers in Kalol villages and large farmers in Dahod villages attempt to be self-sufficient in food and fodder. Also, the market involvement of even large farmers is limited in Dahod villages. The data on marketed surplus suggests that Ghoda and the other Kalol village Sureli are more commercialized than the two Dahod villages. The former also have more productive agriculture. The green revolution technology inputs like irrigation, HYV seeds, chemical fertilizer, pesticides, etc., are also used more intensively in the two villages of Kalol, especially in Ghoda.

Social and demographic characteristics

Tables 4 and 5 present some social and demographic characteristics of cultivator and artisan households respectively.

Table 3: Nature of agriculture in surveyed villages[a], 1981-82

		Villages			
	Items	Ghoda	Sureli	Gangardi	Dadur
1.	Agricultural productivity (Rs./ Acre)[b]	1155.3	695.3	450.5[c]	421.9
2.	Cropping intensity	148.0	148.0	160.5	154.6
3.	GIA as a proportion of GCA (%)	42.1	31.4	16.1	14.9
4.	Proportion of GIA irrigated by wells (%)	93.2	100.0	47.8	88.1
5.	Use of manure (carts/acre)	2.1	1.9	2.1	1.7
6.	Use of chemical fertilizer (Kg./acre)	89.4	29.4	12.8	4.9
7.	Use of pesticides (Rs./acre)	5.7	6.1	1.5	0.9
8.	Cropping pattern: Percentage are under				
	a) Summer *bajri*	27.8	16.0	—	—
	b) Coarse cereals	49.9	51.9	37.0	38.6
	c) Pulses	3.6	2.3	24.9	18.4
	d) Foodgrains	76.9	77.0	99.3	95.7
	e) Groundnut	20.0	19.7	—	1.1
	f) All oilseeds	20.3	21.2	—	1.3
	g) Cotton	0.8	1.7	—	2.8
	h) Non-foodgrains	23.1	22.9	0.7	4.3
9.	Marketed surplus (%)	59.9	34.3	23.7	17.2
10.	Proportion of HYV sowings to total sowings (major crops)	82.0	49.3	50.0	22.8

a   This table is based on the information collected through questionnaires.

b   Value of crop output and by-products per unit area of GCA.

c   In Gangardi, one medium cultivator did not give details about area sown but provided only output figures. Inclusion of this household underestimates cropping intensity and overestimates productivity. This household has been dropped for above estimations.

GIA = Gross Irrigated Area;   GCA = Gross Cropped Area.

Cultivators: Ghoda, the most productive village, had the highest average size of cultivator household, followed by Gangardi, Sureli and Dadur, in that order. The sex ratio (females per 1000 males) of cultivator households was, however, the highest in Dadur at 1060, while the other three villages reported a deficit of females. This peculiar feature can partly be explained by migration of male family members from the village (Dadur) because of lack of work on family farms and non-availability of paid employment within the region.[4]

The cultivator households in Kalol villages, especially Ghoda, mainly belong to high caste groups, while the Dahod cultivators are predominantly scheduled tribes and other backward castes.

The educational status of the sample cultivators showed a very high proportion of illiterates, ranging from about 40 per cent in Ghoda to 71 per cent in Dadur with Gangardi and Sureli falling in between with 46 and 57 per cent respectively. Thus literacy was relatively higher in an agriculturally developed village (Ghoda) and in a business centre (Gangardi). The proportion of graduates and of persons having higher secondary education was also higher in these two villages.

Artisans: Four types of artisans involved in the production of agricultural implements were covered by our survey: carpenters, blacksmiths, leather workers and bamboo workers. Table 5 provides the distribution of sample artisan households by artisan groups and location. All the selected villages except Dadur had blacksmiths and carpenters: there were no blacksmiths in Dadur. While only Dadur had leather workers, bamboo workers were found only in Gangardi.[5]

A review of the origin and history of these artisan households shows that while there is a very long tradition of specialized blacksmithy in all the four villages, only Kalol villages have had access to specialized carpenters. The carpenters in Dahod villages are not carpenters by caste, but have acquired the skill over time (see introduction). Since quite a large proportion of cultivators make their own wooden implements in Dahod villages, the need for specialized carpenters may not have been felt. The implements produced by the cultivators are more functional, less finished, and perhaps less durable. Some households in Sureli, which are not carpenters by caste, have also taken up the profession. As mentioned earlier, apart from the artisans based in the selected villages, three artisans producing improved agricultural implements on a larger scale were also interviewed. One of them, a blacksmith, was based in the village of Garbada in Dahod *taluka* and the other two (one carpenter and one blacksmith) in Kalol town. The latter two artisans had some experience of working in factories before they set up their own workshops, while the Garbada artisan learnt by repairing the improved implements.

The average size of the blacksmith and carpenter households was 5.1 and 7.4 respectively (Table 6). The average size of the other artisan households was somewhat lower (4.5). The proportion of illiterates is higher among carpenter households compared to blacksmith households. The illiteracy rate was even higher among leather and bamboo workers, who were harijans (scheduled caste). None of the family members of the artisan households had received any technical education. This is true of artisan families in Kalol town as well despite the fact that their workshops have highly mechanized fabrication facilities.

Table 4: Social and demographic characteristics of cultivator households, 1981-82

		Villages				
		Ghoda	Sureli	Gangardi	Dadur	All
1.	Household size	8.1	7.9	7.9	7.1	7.7
2.	Sex ratio (Number of females per 1000 males)	930	950	860	1060	943
3.	Education: (both sexes)					
	a. Illiterates (%)	39.5	57.0	46.2	71.3	52.9
	b. Higher secondary (%)	24.1	5.7	18.3	3.5	13.2
	c. Graduates (%)	2.5	0.6	5.7	0.7	2.4
4.	Caste:					
	a. High castes (%)	100.0	65.0	25.0	5.0	48.7
	c. Scheduled castes (%)	—	25.0	—	10.0	8.7
	c. Scheduled tribes and other backward castes (%)	—	10.0	75.0	85.0	42.6

The number of cultivator households surveyed was 20 from each village.

Table 5: Distribution of sample artisan households by artisan groups and villages

Village/town	blacksmiths	carpenters	leather workers	bamboo workers
1. Ghoda	2[a]	—	—	—
2. Sureli	4	2	—	—
3. Gangardi	4	1	—	1
4. Dadur	—	1	1	—
5. Garbadar	1	—	—	—
6. Kalol town	4[b]	1	—	—
All	15	5	1	1

Notes: a Carpenter/blacksmith families.
   b Four families with separate kitchens, whose members jointly own and work in the same workshop.

Economic characteristics

The land and occupational distribution along with some other economic characteristics of cultivator households are presented in Table 7. Table 8 presents the occupational distribution of artisan households.

Cultivators: The distribution of ownership holdings[6] in the selected villages brings out the following characteristic features:

○ The degree of landlessness is significantly higher in the agriculturally most developed village (Ghoda) and in the village where non-agricultural sector provides relatively more employment opportunities (Gangardi, as we shall see later).

○ The share of marginal holdings in total holdings is much higher than their share in area owned. The share of small cultivators in area owned and number of holdings is more or less the same. The share of medium and large farmers in area owned, on the other hand, is much higher than their share in the number of holdings.

○ The average size of land holding in the most productive village (Ghoda) is roughly twice that in other villages. This is primarily because Ghoda is the only village in which medium holdings formed a majority of all holdings and accounted for a major share of the area owned.[7] In other villages the number of medium holdings is very limited although they

21

Table 6: Social and demographic characteristics of artisan households, 1981-2

Characteristics	Blacksmiths	Carpenters	Others[a]	All
1. Sample size	15	5	2	22
2. Household size	5.1	7.4	4.5	5.6
3. Sex ratio (Females per 1000 males)	791	762	800	783
4. Educational status[b]				
a. Illiterates (%)	14.3	35.1	77.8	25.2
b. Higher secondary (%)	24.7	29.7	—	24.4
c. Graduates (%)	1.3	—	—	0.8
5. Caste				
a. Artisan castes[c] (%)	100.0	40.0	—	77.3
b. Other castes (%)	—	20.0	—	4.5
c. Scheduled caste/tribe (%)	—	40.0	100.0	18.2

Notes   a   One is a bamboo worker household and the other is a leather worker.

        b   The percentages under the subgroups a, b and c do not add up to hundred because those who are literate but have not passed their higher secondary examination have not been shown as a separate group.

        c   The artisan caste referred to here is *panchals*. This caste group is traditionally engaged in carpentry and blacksmithy.

own a significant proportion of the area cultivated. The number of holdings large enough to find the adoption of any specific technology economically viable is, therefore, likely to be higher in Ghoda as compared to other villages. This is because the proportion of area under irrigation is much higher in this village than in others. This has obvious implications for the rate and extent of diffusion of those technologies which are not size-neutral.

O There seems to be an inverse relation between agricultural development and the extent of joint cultivation. The holdings under joint cultivation may have a relatively larger family labour supply and therefore may not feel the necessity of labour saving technological change.

Livestock distribution: The average number of bullocks owned per cultivating household was also the highest in Ghoda (Table 7). Moreover, the bullocks per household for all categories of farmers were found only in Ghoda; small

Table 7: Land distribution in surveyed villages and economic characteristics of sample population, cultivator households, 1981-2

Item	Villages			
	Ghoda	Sureli	Gangardi	Dadur
1. Cultivated land per person (acre)	0.9	0.5	0.2	0.5
2. Landless households	29.9	20.6	50.4	3.8
3. Distribution of holdings:				
a. Marginal (0-1 hectares)	3.8 (0.8)	45.4 (17.4)	54.3 (21.2)	35.5 (13.7)
b. Small (1-3 hectares)	43.4 (24.3)	45.4 (48.4)	32.4 (34.0)	47.6 (45.1)
c. Medium (3-7.5 hectares)	50.9 (70.2)	7.8 (22.2)	13.3 (44.8)	16.9 (41.2)
d. Large (7.5 hectares and more)	1.9 (4.6)	1.3 (12.0)	—	—
4. Average size of holding (acres)	8.1	3.7	3.5	4.5
5. Proportion of joint holdings[a] (%)	1.9	5.6	8.6	18.5
6. Worker population ration (Workers per 1000 population)	309	335	316	287
7. Occupational Distribution:				
a. Cultivator	94.0 (2.0)	75.5 (3.8)	58.0 (10.0)	92.7 (—)
b. Agricultural labour	—	13.2	8.0	4.9
c. Non-agricultural labour	—	3.8	6.0	—
d. Regular salaried employee	6.0	7.5	16.0	2.4
e. Traders	—	—	12.0	—
8. Average annual income per cultivator households[b] (Rs.)	15481	7639	8783	3745
9. Per capita annual income (Rs.)	1911	967	1112	524
10 Number of bullocks per household	2.4	1.9	2.0	2.0

Note: 1. For Item 3, figures in parentheses give the proportion of area owned by these categories of farmers.
2. For Item 7a, figures in parentheses give the proportion of cultivator households which cultivate entirely with hired labour.
a Holdings which are jointly owned and cultivated by more than one household, having different kitchens.
b Estimate is based on income from all sources: Crops, animal husbandry, labour, trade, household industry, salary and others.

Table 8: Worker population ratios and occupational distribution of workers in the sample artisan households, 1981-2

Occupational Group	Blacksmiths	Carpenters	Others	All
1. Worker-population ratio (no. of workers per 1000 population)	331	324	222	321
2. Occupational groups (percent of workers)				
a. Artisan services	77.8	58.3	100.0	73.0
	(6.8)	(32.5)	(50.0)	(16.8)
b. Cultivators	17.0	25.0	—	18.6
	(26.1)	(56.7)		(34.1)
c. Agricultural labour	—	—	—	—
		(3.3)	(50.0)	(3.5)
d. Regular salaried employee	5.2	—	—	3.4
e. Traders	—	16.7	—	5.1
	(7.4)			(4.8)

Notes:
1. Figures in parentheses give the proportion of workers reporting the specific occupation as their subsidiary activity.
2. 'Others' include one bamboo worker and one leather worker family.

and marginal cultivators of other villages invariably had fewer than two bullocks per household. The medium and large cultivators, however, had two bullocks each. Also, only Ghoda cultivators have one milch animal per family among all categories of cultivators.

Occupational Distribution: The worker population ratio according to the main activity status of the population of the sample cultivator households was not significantly different among the four villages; between 29 and 34 per cent of the total population of cultivator households formed the workforce. The distribution of workers in cultivator households by their main occupation shows that they spent most of their time cultivating their own or others' plots. For the village as a whole, the proportion of such workers was more than 75 per cent in all villages except Gangardi, where a significant proportion of workers were involved in other activities such as trade, and reported cultivation as their secondary activity. The percentage of cultivators was the highest in Ghoda, agriculturally the most developed village, but was quite high also in the most backward village Dadur. Some members of marginal farmer households reported wage labour as their main occupation, particularly agricultural labour. No member of the cultivator households was involved in wage labour in Ghoda. Regular and salaried employees constituted a very small proportion of total workers. Very few cultivators depended mainly on hired workers to cultivate their land. In Gangardi, however, they constituted about 10 per cent of total cultivator households, because of landowners' involvement in non-agricultural activities. Work as a helper in cultivation was the most important subsidiary occupation followed by agricultural labour in all the villages. Ghoda was the only village where no-one reported agricultural or other labour as his or her secondary occupation.

Levels and sources of income

Ghoda stands out as the most prosperous village in terms of not only agricultural productivity, but also per capita and per household income of cultivator households. The distribution of income by category of farmers and by source reflects the occupational pattern noted above. The cultivator households of Gangardi had the most diverse sources of income while Ghoda cultivators relied exclusively on income from cultivation.

Artisans' occupational distribution: The crude worker population ratios among artisan households by main activity status range between 222 per thousand population to 331; the lowest in 'other' artisan households and the highest among blacksmiths (Table 8). The occupational distribution of workers in artisan households by main and subsidiary occupations shows that income from artisan services is supplemented mainly by income from cultivation. In fact, for some carpenter households of Gangardi and Dadur villages, cultivation was the major source of income. In general, most of the artisans owned some land. A blacksmith in Sureli was even renting out land. The harijan artisans (leather and bamboo workers), however, did not own any land and supplemented their income by agricultural wage labour. The occupational diversity of artisan households may have been more marked than shown in Table 8 since migrant members who had moved out of the households are not covered in the table. In general, very few sons of the present heads of the artisan households (mainly blacksmiths and carpenters) have taken up their fathers' trade.

Conditions of artisan production: The artisans covered in the survey typically worked as individuals with some of the family members, mainly sons, assisting them as and when the need arose. In the surveyed villages, apart from the family members, customers also assist the artisans, particularly blacksmiths. They may operate the bellows or the fan, or beat the hot iron while standing opposite the blacksmith. While the assistants use big hammers, the main blacksmith, who usually sits to work, uses smaller hammers. These village artisans normally work on the verandah of their houses: only the Kalol town artisans have separate worksheds. Artisans take part in all operations from purchasing of raw materials to sales.

Some division of labour was evident only in the workshops of the Kalol town blacksmith. The workshop, jointly owned by four brothers had many power-driven tools. The older members of the family worked on different machines in the workshop: one on the lathe, another on the drill, a third on the metal cutter, and so on. These 'master blacksmiths' occasionally assisted each other but mainly worked on specific machines. This workshop was the only one in the sample which employed workers on a regular basis. The other artisans, if they hired wage labour at all, did so on a daily basis during peak work periods. The Kalol blacksmith also put out some work, unlike other artisan households in the sample. Only the artisans in Garbada and Kalol town used power-operated machines. The other artisans in the sample used manual techniques and a similar range of tools.

A typical village blacksmith has an iron anvil with a variety of 'horn' attachments. The charcoal-fired hearth is placed next to the anvil, and operated by bellows or fan blowers. The Ghoda blacksmiths only used fans, while, in Gangardi and Sureli, both fans and leather bellows were used. Gradually, however manually-operated fan blowers are replacing bellows. Blacksmiths also use various hammers, tongs, mallets, files and chisels. Some also have traditional hand-operated drills to make smaller holes and a vice. They usually make their own files, mallets, tongs, and chisels and horns.

A typical carpenter uses various types of axes, chisels, files, saws, planes, drills, gimlets, screw drivers, vise, right angles etc. Of these, planes and drills are usually home-made. They also make a stool-like device called a *ghodi* on which they sit while cutting wood. Artisans who do both carpentry and blacksmithy as in Ghoda, have both types of tools. The Garbada artisan not only had blacksmithy tools but also a power-operated welder, grinder, blowing fan, and driller.

Artisan workshops in Kalol are more mechanized. The blacksmith had two power-driven lathes, hacksaw, drilling and welding machines, a bench grinder and a power blower. In addition, the workshop had a heavy hand-press, a large vice, and plate cutting and bending machines. The Kalol carpenter had a power-operated drill, wood-turning lathe, wheel-turning lathe and a bandsaw. This workshop also had a large vise. The remaining artisans, leather workers and bamboo workers, are also organized individually, with the help of some family members. while bamboo workers work with a knife alone, the leather workers have a typical cobbler's tool set.

Agents: Apart from artisans and cultivators, the third party involved in the sale or purchase of agricultural implements are the agents who market implements produced by others. Two agents, a private agency in Kalol town and a co-operative in Gangardi, operate in the study region. Both are recognized by the *taluka* and the village Panchayats for their subsidy schemes for purchasing improved implements. These agents sell only those implements which are produced by certain manufacturers recognized by the Gujarat Government.

For many years, only co-operatives were eligible for subsidy for sales of agricultural implements in Gujarat. Any co-operative could order approved implements from the registered manufacturers and arrange to give subsidy to the farmers. In some cases, implements sold by private dealers for a few recognized manufacturers are also eligible for subsidy (Patel, *et al.*, 1982, Chapter IV).

Kisan Fertilizers in Kalol town is a private agency which started dealing in agricultural implements in 1973. Previously, they were dealing only in fertilizers, pesticides, hardware and ropes etc. The spray pump was the first implement they marketed for use with the pesticides they stocked. The agency caters mainly to the villages of Panchmahals district but the market, according to them, is very limited. In Kalol *taluka*, customers mainly come from Derol, Malav, Tarvada, Ghoda, Rabod, Alva, etc. Implements manufactured by Mamta Iron Works, Billimora, and sold by this agency are eligible for government subsidy.

The Gangardi co-operative has been in operation since 1921[8]. At present 30 villages are covered, nine in Dahod *taluka* and 20 in Limkheda *taluka*. Both Gangardi and Dadur, the two surveyed villages in Dahod *taluka* are supplied by this co-operative.[10]

Table 9 gives a brief account of the co-operative's activities during 1981-82 relating to certain aspects of agriculture. Carry-over stocks or inventories are the likely cause of the differences in the value of sales and purchases reported in the table. It can be seen that sale and purchase of agricultural implements constitutes only a small proportion of the co-operative's total outlays; biochemical inputs, HYV seeds and chemical fertilizers are the major agricultural inputs traded.[11] Selling and hiring of agricultural implements is also a relatively new activity for the co-operative, started only in 1977.

The co-operative does however, provide loans for the purchase of agricultural implements. While the short term loans are given mainly for the purchase of seeds and fertilizer, the long-term loans are made for the purchase of engines, bullocks, buffaloes, carts and the digging of wells. All cultivators are eligible for short-term loans, but only cultivators belonging to scheduled tribes can obtain long-term loans along with a 50 per cent subsidy.

On the basis of the village profiles attempted in this chapter, Ghoda can be termed as the most developed of the four study villages. It is followed by Sureli, Gangardi and Dadur in that order. This ranking is based on a limited set of development indicators, including infrastructure, literacy, agricultural productivity, irrigation, use of biochemical inputs and commercialization. The conditions of peasant production and other socio-economic and infrastructural characteristics of the region, reflected in the level of development of the four villages, are presumably related to the use of agricultural implements, to the development and diffusion of agro-mechanical technology and to the

Table 9: Agricultural activities of the Gangardi co-operative, 1981-82

	Item	Purchase (Rs)	Sale (Rs)
1.	Improved seeds	3,45,998.30 (38.87)	3,74,227.22 (43.09)
2.	Chemical fertilizer	4,88,384.87 (54.87)	4,31,652.86 (49.70)
3.	Pesticides	11,438.59 (1.29)	15,632.22 (1.80)
4.	Improved agricultural implements	18,766.65 (2.11)	17,687.25 (2.04)
5.	Cement	25,500.17 (2.86)	29,223.07 (3.37)
6.	Total	8.90,088.58 (100.00)	8,68,422.62 (100.00)

Source: Annual Report: Balance Sheet and Profit and Loss Account (1981-2), Gangardi Vibhag Krushi Sewa Sahkari Mandli Ltd., Gangardi.

conditions of artisans' production. In order to explore these links, an analysis of the market of agricultural implements is also necessary.

Notes

1. Analysis of past data on rainfall shows that the average annual rainfall for Kalol *taluka* is 1052.8mm spread over 41.6 days; for Dahod it is 810.4mm spread over 40.9 days. (Government of Gujarat, *Panchmahals Gazetteer*, 1972, Statement 1.2 pp.58-59).
2. The Dahod settlement report states that: History shows that out of every three years, at least one is bad in both Kharis and Rabi seasons and only one out of every two can be regarded as reasonably satisfactory; *Papers Relating to the Revision Settlement of the Dahod Taluka of the Panchmahals District*, Bombay 1927, p.41, as quoted in, Government of Gujarat, *Panchmahals Gazetteer*, 1972, p.319).
3. See brief accounts of famines and droughts in Panchmahals district for the years 1935-36 to 1968-69 in Government of Gujarat, *Panchmahals Gazetteer*, 1972 pp.321-24.
4. Our field observations show that some male children, between 10 and 18 years of age join their elder brothers or uncles to earn some money and send back a part of it to the family in the village. The female migration, particularly that of children, is usually on a temporary basis.
5. There were five leather worker households in Dadur and four bamboo worker households in Gangardi. These clusters of households were located away from the main village. At the time of survey, members of all the households came together and instead of filling questionnaires for individual households, a group discussion was held. Household level information was collected from only one household from each of the two artisan groups.

28

6. Discussion on land distribution is based on the information collected during the house listing of the selected villages by us. The data, therefore, pertain to the whole village and not only to sample cultivators. In the census house listing preceding the survey, the mortgaging or leasing of land was not reported by any household and hence no difference was envisaged between the distribution of ownership holdings and of operational holdings. But during the sample survey, quite a few of the sample households reported not to be cultivating their own land — they had either mortgaged it out or rented it out, mostly on share cropping basis. This kind of leasing out was more prevalent among small and marginal farmers, particularly in Sureli, the backward village of Kalol *taluka*.

7. We formed the impression during the field survey that the medium cultivators in Ghoda under-reported their land. Some of them were in fact large farmers according to the information provided by other cultivators in the village.

8. Tables on the sources of income by size group of holdings are not reproduced here to save space. The tables bearing on this point are available from ITDG.

9. The co-operative acquired its present name: 'Gangardi Vibhag Mota Kadni Kheti Vishyak Vividh Karyakari Sahkari Mandli Ltd' in 1982.

10. The villages covered in Dahod *taluka* are: Gangardi, Bharsada, Nadwai, Tunkivajee, Tunkianop, Dadur, Bheh, Patia and Nasdwa.

11. Basic consumer items like tea, sugar, foodgrains and cloth are also sold through the various outlets of the co-operatives.

Chapter 3
The Agricultural
Implements Market

Diffusion of an agricultural implement takes place as individual farmers adopt it. The number or proportion of farmers using a particular implement in a region is a measure of diffusion among farms (inter-farm diffusion). Simultaneously, the diffusion of the implement may take place within a farm also, increasing the intensity of the implement's use in the farm (intra-farm diffusion).

A cultivator procures an implement for use on his own farm either by purchasing it — alone or with others — borrowing or hiring it from others. Thus the ownership pattern and the hiring, borrowing and sharing arrangements together determine the extent of use of an implement or its inter-farm diffusion. The actual use of an implement is not, however, covered by these data. To get an estimate of the intensity of use of a specific agricultural implement, we need detailed data on the number of times a cultivator uses it, the crops for which it is used, the area covered, the quantity of output and so on. It is extremely difficult to collect data on all these variables when one is analysing the diffusion of a whole set of agricultural implements. The data presented here indicates only the inter-farm diffusion and not intra-farm diffusion. In later chapters, some aspects of this type of diffusion are also discussed.

Apart from the ownership pattern and hiring, borrowing and sharing practices, the other important dimensions of the agricultural implements market are the sources of supply, prices, and in sales trends, as well as the average duration of useability. High prices of implements generally constrain their adoption. Differentials in prices paid across various cultivator groups reflect imperfections of the market and/or differences in the quality of agricultural implements.

Often the adoption of a new agricultural implement is delayed because of the long life span of the old implement which the new one is intended to replace. Cultivators will not replace old equipment until it is absolutely necessary. On the other hand, more durable substitutes may be preferred by cultivators to less durable ones. In this context it is useful to analyse the age structure of implements in use as it reflects the durability or life span of specific implements, and is therefore indicative of rates of replacement.

We have already noted that the non-availability of fabrication and repair facilities in the vicinity can hamper the diffusion of specific agricultural implements. It is important, therefore, to ascertain the extent to which the farmers' demand for agricultural implements is met locally. An analysis of the sources of agricultural implements owned by the cultivators and what is available locally helps us to identify the gaps in local supply. With this perspective, we can analyse the ownership pattern of agricultural implements, their sources, prices and age structure. This is followed by a discussion of the borrowing, sharing and hiring arrangements prevalent in the study region. Finally, the supply situation of various implements within and outside the villages is examined.

Ownership pattern of agricultural implements

Various agricultural implements are used for different operations in the process of agricultural production. The agricultural implements used in the region and their function are listed in some detail in the introduction. These implements can be broadly classified into four categories according to the source of power used:

Manually operated traditional implements: These implements are small, inexpensive and akin to those which the farmers have been using traditionally. No change has come about in the basic design of these implements except for the substitution of a few wooden parts by iron ones, e.g., rakes (*kadial*). The first six implements in this category are: sickles, hand hoes, axing, digging hoes, pick-axes and crow bars (see Table 10) along with bill hooks (*pariyun*), which can also be classified as tools of agricultural labourers; in some cases, the labourers have to bring these tools with them to work.

Manually operated modern implements: These implements are also manually operated but they are of recent origin, traditionally not known to the farmers. These modern implements are more expensive than the traditional tools above.

Bullock-drawn and related implements: Implements in this category are either directly drawn by bullocks for various operations or used as accessories to these implements. Most of these implements are also traditionally known to the farmers.

Power-driven implements: This category includes tractors and tractor-driven implements, oil engines and electric motors. These implements are very expensive.

Table 10 gives the average number of agricultural implements owned per 100 cultivating households for all villages. The ownership pattern by size

Table 10: Average number of agricultural implements owned per hundred cultivating households (1982)

	Implement	Ghoda	Sureli	Gangardi	Dadur
				Villages	
A: Manually-operated traditional implements					
1.	Sickles	1230	475	445	405
2.	Hand hoes (Shovels)	175	155	125	105
3.	Axes	145	105	85	85
4.	Digging hoes (Spades)	210	100	45	100
5.	Pick-axes	95	—	70	
6.	Crowbars	150	95	60	70
7.	Fork/rake (*kadial*) (W)	—	—	—	75
8.	Fork/rake (*kadial*) (I)	—	—	45	20
9.	Scythe (*dhariyun*)	165	95	60	90
10.	Bill hook (*pariyun*)	110	110		
11.	Rakes (W)	140	170	—	—
12.	Rakes (I)	95	20	10	—
B: Manually-operated modern implements					
13.	Winnowing fans	60	—	—	—
14.	Dusters	10	5	10	—
15.	Sprayers	45	5	5	—
16.	Cotton stalk pullers	65	10	—	—
17.	Hand chaff cutter	10	—	—	—
C: Bullock-drawn and related implements					
18.	Ploughs (W)	110	100	160	175
19.	Ploughs (I)	5	—	15	5
20.	Yokes	310	185	125	125
21.	Yokes for irrigation (*tareleyun*)	—	—	—	75
22.	Two-tine seed-drills	90	45	—	—
23.	Three-tine seed-drills	135	60	5	—
24.	Seeding funnel (*nalichala*)[a] (W)	65	—	5	—
25.	Seeding funnel (*nalichala*) (I)	60	—	—	—
26.	Seeding funnel (*orni*) (W)	75	—	85	120
27.	Seeding funnel (*orni*) (I)	45	—	10	—
28.	Blade harrows (W)	110	70	5	—
29.	Blade harrows (I)	5	5	30	—
30.	Blade hoes (W)	390	220	—	—
31.	Blade hoes (I)	—	—	10	5
32.	Large blade harrow (*karban*)	—	50	—	—
33.	Planks	95	80	70	90
34.	Scrapers (W)	—	—	—	5

		Col1	Col2	Col3	Col4
35.	Scrapers (I)	—	—	—	25
36.	Scoop (*petari*)	—	—	5	—
37.	Bundformers	55	—	—	—
38.	Puddlers (W)	70	25	15	5
39.	Puddlers (I)	—	—	15	10
40.	*moht* (L)	—	—	5	5
41.	*moht* (I)	—	—	5	5
42.	Carts (W)	20	65	20	30
43.	Carts (T)	80	—	5	—
D: Power-driven implements					
44.	Tractors	5	—	5	—
45.	Trolleys	5	—	5	—
46.	Cultivators	5	—	5	—
47.	Oil engines	—	—	10	—
48.	Electric motors	50	10	10	—

Key: W = wooden; I = iron; T = with tyre; L = leather

a *Nalichala* is a seeding funnel with more than one hole, generally three or four.

b *Orni* is a seeding funnel with a single hole.

c *Moht* is an instrument for lifting water from shallow wells. It is traditionally

group of operational holdings is given in Appendix IV. The high level of agricultural development in Ghoda is reflected not only by the wide-spread use of more modern implements (categories B and D), but also by the higher average number of traditional implements owned by cultivating households. The distribution of essential implements such as sickles, ploughs, blade hoes, seed drills, etc., (categories A and C) is more even than that of other implements. [1] Even marginal households own all the essential implements.

On average, every cultivating household in Ghoda owned both a three-tine and a two-tine seed drill. The type of seed-drill used is determined by the crops grown, extent of irrigation, fertilizer use, etc. The fact that small and medium farmers in Ghoda own both types of seed drill, and in fairly large numbers (nearly every household owned at least one), shows that they use specific seed-drills for specific crops, which was not always the case in Sureli. The same holds true of blade hoes and yokes: small and medium farmers owned three to four yokes and blade hoes each, and even marginal farmers owned more than one blade hoe. This suggests that cultivators in Ghoda maintain separate yokes and blade hoes to be used with ploughs and seed drills of different specifications.

The data for Sureli shows that agricultural labourers' tools, along with some bullock-drawn and related implements (e.g., ploughs, yokes, three tine seed-drills and blade hoes), are more equally distributed than modern implements.[2] The fact remains that not all marginal farmers own ploughs; the extent of seed-drill ownership among them is also limited. The relatively large number of blade hoes per cultivating household (particularly among the medium and large farmers) can partly be explained by the necessity to have different-sized blade hoes for different sized seed-drills. Similarly, yokes of different size are also needed for different-sized blade hoes and for ploughing. Conspicuous by their lack of use are implements in categories B and D — new handtools and power-driven implements. Puddlers and bund formers are also not used.

Gangardi provides a picture similar to that of Sureli. There are however, a few differences:

○ traditional fork rakes (*kadials*) are used instead of rakes in the dehusking process and in the fields spades are used instead of rakes for bunding;

○ two tine and three tine seed-drills are not used in the village and consequently the use of seeding funnels with more than one hole (*nalichalas*) and blade hoes is non-existent. Single-hole funnels (ornis) are tied instead to the ploughs for sowing, and ploughs are used by a few cultivators for inter-cultivation as well;

○ blade harrows are also not common here though some larger farmers have bought iron blade harrows with subsidy;

○ the number of ploughs per cultivator-family is high because of the rocky terrain which increases the wear and tear of the implements used, and farmers have to keep more than one plough for emergencies;

○ bullock-drawn leather/iron *mohts* are still used for irrigation in the village;

○ the use of tractor-drawn cultivators and trolleys was reported by some of the sample farmers.

Dadur again presents a picture similar to Gangardi. Use of *mohts* is, however, much more prevalent in Dadur compared to Gangardi; and scrapers are used only in this village. While in Gangardi most of the cultivators use a slightly improved fork rake (*kadial* with iron rod), cultivators in Dadur still use the traditional bamboo version of the implement. Besides, none of the sample cultivators in Dadur reported ownership of implements belonging to categories B or D.

In general, the patterns of ownership of implements in the four selected villages show that most of the traditional handtools and bullock-drawn implements (categories A and C) are more equally distributed across different sized holdings than modern implements. More specifically, harvesting implements (sickles), general-purpose low-cost manually operated implements, ploughs and yokes are most widely distributed in all the villages. These are followed by sowing implements, i.e., two/three tine seed-drills along with seeding funnels in Kalol villages, and single-hole funnels (*orni*) and ploughs in Dahod villages. For Kalol villages, blade hoes — the bullock-drawn intercultivation implements — are also more widely distributed. The other point which emerges is that the new and improved implements are used only in Ghoda. But even there, the use of bullock-drawn iron implements (iron ploughs, blade harrows and hoes) is not widespread.

In Gangardi and Dadur, some cultivators have bought bullock-drawn iron implements with subsidy but, as we shall see later, they do not use them.

Lastly, a point about the extent and nature of tractorization may be noted. The ownership of tractors is quite limited and as expected, is concentrated in relatively large land holdings, Besides, cultivators who own tractors only keep trailers and cultivators as accessories. This implies that tractors are mainly used to mechanize tillage and transport operations.

Sources of agricultural implements

All four villages have peculiar features of their own as far as sources of agricultural implements are concerned. In general, one finds that Dahod villages have more diversified sources compared to Kalol villages. This diversification partly reflects the inadequate supply facilities for specific implements in these villages. Ghoda cultivators have major market links with the nearby town, Kalol, for implement purchases. Village artisans are the most important source of implements in Sureli, followed by Gangardi. In Dadur, a significant proportion of implements are made at home and by relatives, neighbours or friends; this is true, to an extent, for Gangardi as well. Since Dadur does not have blacksmiths, artisans in the nearby village (Gangardi) have been the other important source of implements.

There are very few instances of a shift from traditional sources (local artisans) to other sources in nearby towns in Kalol villages. Outside sources are used mainly for new implements which cannot be fabricated within the village, i.e., iron implements, and implements in categories B and D. The Ghoda cultivators' market links with Kalol town, however, are of long standing; even bullock-drawn and traditional handtools, normally purchased within the village are bought here. Bullock-drawn implements are procured from Kalol artisans and handtools from other shops in town. One of the shops there has started selling spades, shovels and pick-axes manufactured by Tata Iron and Steel Co. The source has remained the same, but the brand has changed.

There is a shift from traditional sources in Dahod villages due to a weekly fair held in Garbada village during the last five years. Some cultivators in Gangardi and Dadur have started buying a few manually operated traditional implements from this fair instead of buying them from village artisans. Most of the cultivators' needs are satisfied within the village in Sureli and Gangardi, in the nearby village in Dadur and in the nearby town in Ghoda. Besides a significant proportion of cultivators make some of the implements on their own in the Dahod villages, especially in Dadur.[3]

Prices and age structure of agricultural implements

The implements differ in size, weight, and quality of raw material and parts across cultivator categories and villages. It is very difficult to isolate these differences. The blacksmiths in Gangardi, for example, insisted that they use better quality steel and iron than the Kalol artisans. The issue is further complicated by the fact that in some cases raw materials and parts are provided by customers themselves while in others, the artisan provides everything. If the customer provides wood which is cut from a tree in his own field it is very difficult for him to estimate a price for that. In cases where the implements are made by the users themselves, as in Dahod villages, the costs become all the more difficult to ascertain. To overcome some of these problems, we have distinguished between two sets of prices: total prices including all costs and prices excluding the cost of wood. The latter set of prices includes those cases where the respondent could not calculate the value of raw materials provided by him (see Table 11).[4]

The data on the age of agricultural implements (Table 12) also have some limitations. The question which was asked in the survey was: when did you buy this implement? The answer referred to the year when the whole implement was purchased. Some parts of agricultural implements are replaced more often than others. For example, in ploughs, the iron and wooden parts of the ploughshare, which till the land, are replaced more often. It is likely, therefore, that quite a few parts in the implement are new while the implement itself is relatively old. The reported year of purchase or the age of the implement may, therefore, be misleading. To collect information about every part of an implement is very difficult.[5] The data on prices and the age of implements should, therefore, be interpreted with the above limitations in mind.

The variations in reported current prices of manually operated traditional implements across villages are not marked. The notable exceptions are: sickles, shovels and pick-axes (Table 11). The price of sickles is higher in Dahod villages and that of the other two is higher in Kalol villages. The sickle in Dahod villages is a bigger, heavier and non-serrated version of the sickle in Kalol villages. This might be one of the reasons for the higher sickle prices in Dahod villages. On the other hand, the cultivators in Kalol villages, especially in Ghoda, use shovels and pick-axes made by Tata Iron and Steel Co., which are costlier, having an ISI mark, while the Dahod cultivators use locally produced ones which are less costly. The reported price of spades is much lower in Sureli than in other villages, presumably also because of differences in quality. As for modern manually operated implements, in most cases we did not have enough observations to generate reliable price estimates. Table 11 does not, therefore, report these estimates. In general, however, the cultivators in Ghoda and Sureli seemed to use costlier versions of these implements. As noted above, the use of these implements is more widespread in Kalol villages than Dahod villages.

For bullock-drawn implements also, the reported current prices are higher in the Kalol villages than in the Dahod villages. Between the two Kalol villages, the prices are generally higher in Ghoda than in Sureli. A similar situation exists in Gangardi compared to Dadur in Dahod. In general, one finds that these implements are of higher quality in Kalol villages (especially

Table 11: Average current price[a] of agricultural implements (Rs.) (1982)

		Villages			
Implement		Ghoda	Sureli	Gangardi	Dadur
1		2	3	4	5
A: Manually-operated traditional implements					
1.	Sickles	4.8	5.1	6.0	6.3
2.	Hand hoes (Shovels)	21.8	18.1	14.7	14.5
3.	Axes	19.5	21.2	17.7	21.2
4.	Digging hoes (Spades)	19.9	14.8	19.1	23.8
5.	Pick-axes	30.8	—	18.8	
6.	Crowbars	21.0	16.8	18.8	19.0
7.	Fork/rake (*kadial*) (W)	—	—	—	2.7
8.	Fork/rake (*kadial*) (I)	—	—	6.2 (3.0)	4.5[b]
9.	Scythe (*dhariyun*)	20.7 (17.0)	23.8	14.2	15.8
10.	Bill hook (*pariyun*)	16.7	10.4		
11.	Rakes (W)	11.4 (3.0)	15.0 (5.9)	—	—
12.	Rakes (I)	14.9	8.5	—	—
B: Manually-operated modern implements					
13.	Winnowing fans	367.5	—	—	—
14.	Sprayers	500.0	—	—	—
15.	Cotton stalk pullers	34.2	—	—	—
C: Bullock-drawn and related implements					
16.	Ploughs (W)	115.6 (67.5)	71.4 (46.9)	47.3 (19.2)	38.7 (9.7)
17.	Yokes	36.5	30.9 (10.0)	17.1 (6.0)	15.64 (10.0)
18.	Yokes for irrigation (*tareleyun*)	—	—	—	—
19.	Two-tine seed-drill	81.2	65.7 (35.0)	50.0	—
20.	Three-tine seed-drill	116.3 (75.5)	81.4 (50.0)	—	—
21.	Seeding funnel (*nalichala*)[a] (W)	49.4	—	—	—
22.	Seeding funnel (*nalichala*) (I)	84.0	—	—	—
23.	Seeding funnel (*orni*) (W)	7.2	—	4.7	4.7
24.	Seeding funnel (*orni*) (I)	22.7	—	—	—

25.	Blade harrows (W)	144.0 (82.5)	105.0 (68.7)	—	—
26.	Blade harrows (I)	—	—	42.5	—
27.	Blade hoes (W)	52.9	40.0 (36.7)	—	—
28.	Large blade harrow (*karban*)	—	86.7 (47.5)	—	—
29.	Planks	48.3	26.9 (15.7)	65.0 22.5)	30.3
30.	Scrapers (I)	—	—	—	48.3[b]
31.	Bundformers	97.9 (46.0)	—	—	—
32.	Puddlers (W)	62.5 (12.0)	—	—	—
33.	*moht* (L)	—	—	—	164.2
34.	Carts (W)	2575.0[b]	1757.1 (1150.0)	2250.0[b]	2175.0
35.	Carts (T)	3300.0	—	—	—

Key: I = iron; T = with tyres; W = wooden; L = leather; NR = not reported

Notes: a The estimates of average current prices of some implements are not reported because of small number of observations (less than 3)

b Cases where the number of observations was still small (3-5) making the estimates reliable.

c Figures in parentheses indicate prices which do not include the cost of wood and, in some cases, other raw materials.

Ghoda) than Dahod villages. Also, in Kalol villages, all these implements are made by the artisans while in Dahod villages quite a number are made at home and are crude in nature.[6]

Table 12 provides the average age of the implements owned by sample cultivators.[7] The manually operated traditional implements are, in general, older in the Dahod village than in Kalol villages. The other manually operated implements, which are mostly used in Kalol villages, are older in Ghoda than Sureli. This is partly because all these implements have a relatively long life and their use was initiated earlier in Ghoda. Very few bullock-drawn implements are common to Dahod and Kalol villages. Among the comparable ones, the average age of ploughs and seeding funnels (*orni*) is lower in the Dahod villages than in the Kalol villages. This is mainly for two reasons: the land in Dadur and Gangardi is very stony and undulating so that the wear and tear of ploughs and other similar implements is very much higher; and the quality of these implements is much better in the Kalol villages. The higher prices of most implements in the Kalol villages than in the Dahod villages reflect marked quality differences. Some cultivators in the Dahod villages argued that there is no point in making more expensive and sophisticated wooden ploughs because in cases of breakage the loss is very high.

Borrowing, sharing and hiring arrangements

The purchase of agricultural implements entails a fixed cost. While the agricultural implement itself is not divisible, its use can be restricted to specific crops or area. The practice of hiring, borrowing and sharing, to an extent, reduces the indivisibility of agro-mechanical technology for an individual farmer as he can use the implement without owning it. This obviously helps the diffusion of that implement. Various arrangements of borrowing, sharing and hiring of implements were prevalent in the four study villages. In general, neighbours and relatives borrow implements from each other. The practice of sharing is mainly restricted to close relatives such as brothers, father and son, etc. The implements are hired either from fellow cultivators, the Panchayat or the co-operative.

Borrowing and sharing: Borrowing and sharing arrangements were found to be more prevalent in the Kalol villages than in the Dahod villages. In Ghoda, the most developed village, the manually operated modern implements (category B) were borrowed and shared along with some new bullock-drawn implements such as bundformers. The hiring of some manually operated modern implements, e.g. dusters and sprayers, was prevalent in Ghoda earlier but is almost non-existent now.[8] Mainly small and medium farmers were involved in these transactions. Marginal farmers in the sample also borrowed bullock-drawn bundformers and puddlers. In the backward villages (Sureli, Gangardi and Dadur), agricultural labourers' tools, bullocks and some bullock-drawn traditional implements were borrowed and shared.[9] These transactions took place mainly among marginal farmers.[10] In Sureli, where this system of borrowing and sharing of these implements is most prevalent, some small cultivators also borrowed and shared implements. But in these cases, bullock-drawn implements such as seed-drills and blade hoes were involved. Medium and large farmers do not participate in these arrangements as far as implements of category A and C are concerned. Some farmers (other than marginal) in the Dahod villages also borrow category B implements from the Panchayat or the co-operative, mainly plant-protection equipment such as dusters, sprayers etc. The borrowing of iron bullock-drawn implements like ploughs and blade harrows was also reported in Ghoda and Gangardi to a limited extent.

Hiring: Tractors (for tillage, transport and threshing), diesel engines, electric motors and threshers were the main items hired in the sample villages. Hiring of these implements was most widespread in Ghoda; in other villages, especially in the Dahod villages, it was limited. Unlike in Ghoda, plant-protection equipment was also hired in Sureli and, to some extent, in

Table 12: Average age of the agricultural implements owned (years, 1982)

	Implement	Villages			
		Ghoda	Sureli	Gangardi	Dadur
	1	2	3	4	5
A: Manually operated traditional implements					
1.	Sickles	1.4	1.2	1.8	2.2
2.	Hand hoes (Shovels)	4.0	2.7	5.1	3.7
3.	Axes	3.6	4.1	6.4	6.4
4.	Digging hoes (Spades)	3.4	3.1	5.0	4.4
5.	Pick axes	5.1	—[a]	6.0	
6.	Crowbars	7.8	3.6	5.1	7.6
7.	Fork/Rake (*kadial*) (W)	—	—	—	0.5
8.	Fork/Rake (*kadial*) (I)	—	—	4.4 (5.5)	2.6[b]
9.	Scythe (*dhariyun*)	3.8 (4.0)	4.5	7.0	4.0
10.	Billhook (*pariyun*)	2.2	2.1		
11.	Rakes (W)	2.8 (0.5)	1.0 (1.4)	—	—
12.	Rakes (I)	4.1	3.5**	—	—
B: Manually operated modern implements					
13.	Winnowing fans	2.8	—	—	—
14.	Dusters	5.0*	2.0*	1.0*	—
15.	Sprayers	7.0	5.0*	—	—
16.	Cotton stalk pullers	7.1	2.0*	—	—
17.	Hand chaff cutter	0.0*	—	—	—
C: Bullock-drawn and related implements					
18.	Ploughs (W)	2.8 (3.5)	2.3 (3.6)	1.8 (0.7)	1.7[d] (1.1)
19.	Yokes	4.5	3.9 (6.5)	6.6 (3.0)	3.5 (15.0)
20.	Yokes for irrigation (*tareleyun*)	—	—	—	5.7
21.	Two tine seed-drill	3.7	1.4 (2.7)	4.0	—
22.	Three tine seed-drill	2.7 (2.4)	2.3 (5.0)	—	—
23.	Seed funnel (*nalichala*) (W)	5.3	—	—	—
24.	Seeding funnel (*nalichala*) (I)	4.4	—	—	—
25.	Seeding funnel (*orni*) (W)	4.8	—	1.2	2.0
26.	Seeding funnel (*orni*) (I)	3.4	—	—	—

27.	Blade harrows (W)	3.5 (1.7)	3.0 (4.3)	—	—
28.	Blade harrows (I)	—	—	4.0	—
29.	Blade hoes (W)	3.4	0.0 (1.8)	—	—
30.	Large blade harrow (*karban*)	—	6.3 (4.7)	—	—
31.	Planks	2.0 (5.0)	1.3 (3.0)	6.5 (5.2)	3.8 (1.0)
32.	Scrapers (I)	—	—	—	7.5**
33.	Bundformers	2.6 (1.0)	—	—	—
34.	Puddlers (W)	2.4 (4.0)	—	—	—
35.	Puddlers (I)	—	—	—	—
36.	*moht* (L)	—	—	—	3.6 (3.0)
37.	Carts (W)	5.5**	5.8 (5.0)	6.0** (6.0	7.0 (10.0)
38.	Carts (T)	3.7	—	—	—
39.	Electric motors	0.5	—	—	—

W = wooden; I = iron; T = with tyre; L = leather.

Notes:
   a   The estimates of average age of some agricultural implements are not reported because of the small number of observations (less than three).

   b   An asterix identifies those cases where the estimates are less reliable because of small number of observations, * is used for those cases where the number of observations is less than thtee but more than one and ** denotes those cases where the number of observations is in the range of 3-5.

   c   Age of implements for which the customers provided wood and other raw materials is given in parenthesis.

   d   If we exclude one observation the estimate drops down to 0.40.

the Dahod villages where marginal farmers hired them while others borrowed them from the Panchayat or the co-operative. While marginal farmers participated predominantly in borrowing/sharing arrangements, the small, medium and large farmers predominated in hiring transactions, particularly in Ghoda, Sureli and Gangardi. Apart from medium and large farmers who hire out these services, the co-operative, panchayats and non-cultivating individuals are also involved. While the Gangardi Co-operative hired out both plant-protection equipment and tractor services in Gangardi and Dadur, the Panchayat only provided the former and a non-cultivating individual only the latter. A few large cultivators who have their fields in and around Sureli have formed a co-operative. They have bought a motor and dug a tubewell with a bank loan. They shared the repayment costs and the irrigation facilities. This group is limited to only a few members of the co-operative.

The patterns of hiring, borrowing and sharing observed so far show that the involvement of marginal farmers in such arrangements is limited, particularly in the backward villages. The fact that their involvement is more limited in the case of modern and improved implements as compared to traditional ones suggests that marginal cultivators enter into hiring and other arrangements primarily to complete the basic set of traditional agricultural implements and somehow cultivate their fields. The motives of other cultivators, especially medium and large, who enter into such arrangements are presumably different. A few cases of hiring of implements lend further support to this hypothesis:

○ A marginal farmer in Sureli reported hiring a two tine seed-drill for Rs5 per day.
○ Two marginal farmers in Gangardi hire bullocks along with ploughs and a seeding funnel (*orni*) to till and sow their land. They pay Rs15 to 20 per day.
○ Hiring of bullocks from other cultivators on a long-term basis was reported by a marginal farmer in Dadur. The owner is paid Rs25 per month per bullock. The hirer uses the bullock in his farm and takes care of its fodder needs.
○ A large farmer in Sureli and a small cultivator in Gangardi hired the service of marginal farmers along with their bullocks, ploughs and seeding funnels (*orni*). Both these cultivators were *brahmins*, who did not touch the plough and paid Rs15 per day plus meals to the workers, and provided fodder for the bullocks.[11] However, while the small farmer usually cultivated his land in this manner, the larger cultivator resorted to such hiring arrangements only during peak seasons when his permanent farm servants and hired workers proved to be inadequate for the task.

Thus the marginal farmers enter into hiring arrangements because they do not have even the basic implements; the Gangardi small farmer, above, does so because of caste taboos. The Sureli cultivator, in addition to being a *brahmin*, has a large farm and consequently has to face higher and sharper peaks in labour use. While the Gangardi farmers have the option of hiring tractor services, the Sureli cultivators do not have this facility.

The hiring charges for various implements are fairly uniform across various cultivator categories. Table 13, therefore, does not report them separately for different-sized holdings. Deviations, however, are mentioned as notes to the table. It can be seen from Table 13 that while the services of electric motors and oil-engines are cheaper in the Kalol villages, tractor services[12] are relatively cheaper in the Dahod villages. Oil engine rates are higher in Dadur than Gangardi. Only the Gangardi cultivators used tractor services for transporting while only the Ghoda cultivators used them for threshing and harvesting. In general, as mentioned above, hiring charges are the same for all cultivators but where differential rates exist, the small and marginal cultivators tend to pay a higher price (see Table 13, notes).

Table 13: Hiring charges for various agricultural implements — (Rs, 1982)

Implement	Village			
	Sureli	Ghoda	Gangardi	Dadur
1. Dusters	—	—	2 per day[d]	—
2. Sprayers	1.5-2 or 5 per day[a]	—	2 per day[d]	—
3. Oil engine	7-12 per hr.	10 per hr.	15 and 12 per hr. [e]	15-20 per hr.
4. Electric motor	6-12 per hr.	6-10 per hr.	—	—
5. Tractor				
- tillage	—	70 per acre[b]	40-50 per acre	50 per hr.[e]
- threshing	—	2-2.5 per 20 kg of grain	—	—
- transport	—	—	2 per quintal	—
- harvesting	—	70 per hr.[b]	—	—
6. Threshers	1.5-2 per 20kg of grain	2-2.5 per 20kg of grain[c]	—	—

Notes:
   a   Rs 1.50-2.00 paid by small and medium farmers — they hired either from the Panchayat or from large farmers. Rs. 5/- was paid by one marginal farmer and one small farmer who hired the sprayer from medium farmers.

   b   This rate was paid by most cultivators. The range, however was Rs 60-75/acre.

   c   Only one marginal farmer and one small farmer reported paying Rs 3/- per 20kg.

   d   Only marginal farmers hired, others borrowed.

   e   The marginal and medium farmers paid Rs15 while large ones paid Rs12.

   f   Only one cultivator (small) reported paying Rs 30 per hour.

It was seen in an earlier section that the manually operated traditional tools and bullock-drawn implements were more equally distributed among various categories of cultivators than other implements. Marginal and small farmers did not have adequate sets of these implements. Taken along with the patterns observed in the sharing, borrowing and hiring arrangements, the reported ownership of implements indicates that while the small and marginal farmers broadly aim at completing their set of implements in categories A and C, the medium and the large cultivators complement them by acquiring the services of implements in categories B and D. Ghoda is the only exception where even small and marginal cultivators attempt to acquire the services of all types of implements through borrowing, sharing and hiring arrangements.

Juxtaposing the patterns of agricultural development observed earlier with the patterns of agro-mechanical technology diffusion observed in this chapter, one finds a close association between the two. In general, villages with higher agricultural productivity (i.e., the Kalol villages) use more varied and sophisticated agricultural implements. The extent of adoption of these implements is, however, fairly limited even in the developed villages. The reasons underlying the limited diffusion of these implements will be discussed in some detail in the next chapter. As we shall see, inadequate supply of fabrication and repair services is one of the major reasons for limited diffusion of modern implements.

Sources of supply

Earlier discussion on sources of implements based on information collected from the cultivators indicated that (i) local village artisans primarily produce traditional handtools and some bullock-drawn implements (mainly from categories A and C); and (ii) new handtools, plant protection equipment, bullock-drawn iron implements and power driven equipment (categories B and D) are supplied by large-scale artisans or their agents. In what follows, we examine details of the sources of supply of agricultural implements using information elicited directly from some of the artisans and agents.

Tables 14 and 15 report, for blacksmiths and carpenters respectively, their scale of production and the sources of raw materials and parts. The estimates of average number sold in the agricultural years 1981-2 for each implement are simple averages of estimates provided by the sample blacksmith and carpenters. In the case of Garbada and Kalol town artisans, the estimates are based on single responses. The small number of observations makes the estimates less reliable. Conclusions drawn from these estimates can only be tentative.

Agricultural implements fabricated and sold in the study region: Table 14 suggests that the village blacksmiths normally make traditional handtools (category A) along with some parts of bullock-drawn implements. These parts include the blades of blade harrows and hoes, iron plough shares, iron teeth for puddlers and some iron parts used in bullock carts such as axles, linchpins, metal washer-like discs for the wheel and metal hoops that fit around the wheel. The village carpenters, on the other hand, concentrate on bullock-drawn traditional implements, such as ploughs, seed-drills, harrows, hoes and yokes, as well as carts (see Table 15). Because of lack of demand, Dahod carpenters make a very limited set of implements — mainly ploughs and yokes. The three large fabrication units in Kalol town and Garbada, however, produce a larger set of implements than the other village artisans. They fabricate both the traditional and the new and improved elements. The blacksmith workshop in Kalol town is an exception; only specialized implements such as winnowing fans, some new iron implements (bullock-drawn) and threshers are produced.

The Garbada blacksmith continues to make traditional handtools along with bullock-drawn iron implements. Similarly, the Kalol carpenter continues to make traditional bullock-drawn wooden implements along with modern implements like bund-formers. The implements produced by the Dadur leather worker and the Gangardi bamboo worker are mainly the leather *moht* and the bamboo seeding funnel (*orni*), respectively. Of the agents, the Gangardi co-operative sells mainly plant-protection equipment and bullock-drawn iron implements and Kisan Fertilizers (Kalol) also sells traditional handtools such as spades, pickaxes and digging hoes manufactured by Tata Iron and Steel, among others (see Table 16). The number of agricultural implements sold by the latter is also much higher than that sold by the Gangardi Co-operative.

Market conditions faced by the artisans of the region

Agricultural development, and the associated commercialization, exposes the village to wider market forces. If village artisans are not able to compete successfully with articles produced cheaply on a large scale in towns or by larger fabrication units in rural areas, they are displaced. The manner in which local artisans respond to changing market conditions determines to a great extent the supply of fabrication and repair services which in turn may influence the rate of technology diffusion.

The scale of operation is very uneven across villages and across types of artisans. The fabricators and agents also claimed that demand is uncertain and unstable and year-to-year fluctuations in sales are sizeable. Our data also show that among village blacksmiths, only those in Sureli seem to be assured of a somewhat stable market. The average annual sales of sickles, the most important handtool in terms of sales, is highest in Sureli. Sureli blacksmiths are in fact very famous in the region, especially for their sickles. The blacksmith in Ghoda seem to face the worst situation, but they are blacksmith carpenters and the loss of blacksmithy work may not mean a total loss of work for them. Even the Kalol blacksmith's workshop operates on a small scale but for here, fabricating agricultural implements is only one of several activities. They also take on piece-work from outside.[14] Among the carpenters, the Kalol town artisan is the only one operating on a somewhat larger scale, followed by the carpenter in Dadur.[15] For the Kalol carpenter, making agricultural implements is the main activity, though he does some outside jobs such as cutting wood and making windows and doors.

The market for village artisans is restricted to neighbouring villages. Sureli blacksmiths, well-known in the region, are again an exception. However, the Garbada and Kalol artisans and agents have their market spread over many villages. Thus, in general, the artisans seem to produce for a very small and unstable market. In most cases, the products fabricated by them are made to order and in quite a few cases customers provide raw materials such as wood and iron parts (see Tables 15 and 16).[16] This was true even in the case of the carpenter in Kalol whose scale of production is relatively large. Only

Table 14: Scale of production and sources of raw material and parts, blacksmiths (1981-2)

	Name of the village/town					
	Ghoda			Sureli		
Implement	Average number sold	Source of raw material[d] (code)	Parts[e]	Average number sold	Source of raw material[d] (code)	Parts[e]
1	2	3	4	5	6	7
1. Sickles	100	1	1	283	5	½
2. Axes	—	—	—	10	5	5
3. Digging hoes	—	—	—	10	5	5
4. Scythe (*dhariyun*)	5	1	1	22	5	1
5. Bill hook (*pariyun*)	6	1	1	83	5	1
6. Rakes (I)	—	—	—	—	—	—
7. Winnowing fab	—	—	—	—	—	—
8. Cotton stalk puller	—	—	—	—	—	—
9. Large blades (*pas*)[a]	—	—	—	40	5	—
10. Small blades (*pasiya*)[b]	—	—	—	25	5	—
11. *Dadha* [c]	—	—	—	50	5	—
12. Iron plough	—	—	—	—	—	—
13. Blade harrow (I)	—	—	—	—	—	—
14. Blade hoe (I)	—	—	—	—	—	—
15. Puddler (I)	—	—	—	—	—	—
16. Two tine seed-drill (I)	—	—	—	—	—	—
17. Seeding funnel (*nalichala*) (I)	—	—	—	—	—	—
18. Seeding funnel (*orni*) (I)	—	—	—	—	—	—
19. *Moht* (I)	—	—	—	—	—	—

	village/town						
	Gangardi			Garbada[f]		Kalol Town[f]	
	Average number sold	Source of raw material[d] (code)	Parts[e]	Average number sold	Source of raw material[d] (code)	Average number cold	Source of raw material[d] (code)
1	8	9	10	11	12	13	14
1. Sickles	57	1/4/5	1	50	5	—	—
2. Axes	15	1	1	30	5	—	—
3. Digging hoes	—	—	—	20	5	—	—
4. Scythe (*dhariyun*)	15	1/4/5	1	20	5	—	—
5. Bill hook (*pariyun*)	15	4/5	1	10	5	—	—
6. Rakes (I)	2	3/5	1	15	5	—	—
7. Winnowing fan	—	—	—	—	—	2	3/5
8. Cotton stalk puller	—	—	—	5	5	—	—
9. Large blades (*pas*)[a]	10	4/5	—	—	—	—	—
10. Small blades (*pasiya*)[b]	10	4/5	—	—	—	—	—
11. *Dadha*[c]	—	—	—	—	—	—	—
12. Iron plough	—	—	—	10	5	3	3/5
13. Blade harrow (I)	—	—	—	10	5	3	3/5
14. Blade hoe (I)	—	—	—	10	5	—	—
15. Puddler (I)	2	1	—	50	5	—	—
16. Two tine seed-drill (I)	—	—	—	—	—	—	—
17. Seeding funnel (*nalichala*) (I)	—	—	—	—	—	10	3/5
18. Seeding funnel (*orni*) (I)	—	—	—	30	5	60	3/5
19. *Moht* (I)	—	—	—	30	5	—	—

Source Codes: 1. Customer 2. Village artisan 3. Same village/town 4. Nearby village 5. Nearby town

Notes:
a   Blade for blade-harrow
b   blade for blade-hoe
c   An iron tube which is fixed in the head piece of the blade harrow/hoe and in which the blade is fixed
d   Raw material is mainly iron, also iron sheets for some implements like *nalichala*, *orni*, etc.
e   Wooden handles
f   Estimates based on single responses. The Garbada and Kalol town artisans did not provide details about sources of parts.

Table 14 (continued)

Table 15: Scale of production and sources of raw material and parts, carpenters (1981-2)

	Name of the village/town					
Implement	Ghoda			Sureli		
	Average number sold	Source of raw material[a] (code)	Parts	Average number sold	Source of raw material[a] (code)	Parts
1	2	3	4	5	6	7
1. Rakes	8	1	1[b]	—	—	—
2. Plough	—	1	1[c]	7	1/4	1[c]
3. Blade harrow	—	1	1[d]	1	1/4	1[d]
4. Large blade harrow (*karban*)	—	—	—	1	1/4	1[d]
5. Blade hoe	15	1	1[d]	6	1/4	1
6. Three tine seed-drill	3	1	1[e]	4	1/4	1[e]
7. Two tine seed-drill	4	1	1[e]	1	1/4	1[e]
8. Yokes	10	1	—	3	1/4	—
9. Puddlers	2	1	—	1	1/4	—
10. Cart	—	—	—	1	1/4/5	1/5
11. Planks	8	1	1[f]	—	—	—
12. Bundformer	—	—	—	—	—	—
13. Yokes for Irrigation (*tareleyun*)	—	—	—	—	—	—

	Name of the village/town					
	Gangardi			Dadur[c]		
Implement	Average number sold	Source of raw material[a] (code)	Parts	Average number sold	Source of raw material[a] (code)	Parts
1	8	9	10	11	12	13
1. Rake	—	—	—	5	1	1[b]
2. Plough	7	1	1[c]	25	1	1
3. Blade harrow	—	—	—	—	—	—
4. Large blade harrow (*karban*)	—	—	—	—	—	—
5. Blade hoe	—	—	—	—	—	—
6. Three tine seed-drill	—	—	—	—	—	—
7. Two tine seed-drill	—	—	—	—	—	—
8. Yokes	3	1	—	10	1	—
9. Puddlers	—	—	—	5	1	1[b]
10. Cart	—	—	—	—	—	—
11. Planks	—	—	—	—	1	1[f]
12. Bundformer	—	—	—	—	—	—
13. Yokes for Irrigation (*tareleyun*)	—	—	—	8	1	—

Table 15 (continued)

	Name of the village/town		
	Kalol Town[h]		
Implement	Average number sold	Source of raw material[a] (code)	Parts
1	14	15	16
1. Rake	75	1/3	—
2. Plough	25	1/3	3[c]
3. Blade harrow	38	1/3	3[d]
4. Large blade harrow (*karban*)	—	—	—
5. Blade hoe	56	1/3	3[d]
6. Three tine seed-drill	25	1/3	3[e]
7. Two tine seed-drill	10	1/3	3[e]
8. Yokes	25	1/3	—
9. Puddlers	—	—	—
10. Cart	9	1/3	3
11. Planks	8	1/3	3[f]
12. Bundformer	6	1/3	3[g]
13. Yokes for Irrigation (*tareleyun*)	—	1/3	—

Source Codes: 1. Customer 2. Village artisan 3. Same village 4. Nearby village 5. Nearby town

Notes:
a Raw material is normally wood
b Wooden handles
c Iron plough share
d Iron blades
e Iron shares to be put in tines
f Iron rings
g Thin iron strip and iron rings
h Based on single response

Table 15 (continued)

blacksmiths in Sureli sometimes keep ready-made sickles for sale as their market is spread over a large number of villages. Even the larger fabricators make implements to order. This type of production poses no distribution or marketing problems. As such, the sales procedure may be a function of the size and transportability of products to be sold. However, many small implements fabricated by carpenters and blacksmiths (especially the latter) can be carried around and sold. But only the Gangardi bamboo worker and Dadur leather worker reported that they make implements in large numbers and move around in different villages to sell their products if they happen to get the raw material (leather/bamboo) in bulk.[17] The Garbada blacksmith also at times goes to large cultivators with samples of his iron implements.

One may recall in this context that most of the artisans are also cultivators. While the Sureli artisans either give their land on crop share or cultivate it with hired labour, the artisans in Ghoda and the Dahod villages normally cultivate it themselves with their family labour. This is true particularly of carpenters in the Dahod villages. Thus, a large section of artisan population, especially in the Dahod villages, essentially supplement their farm incomes with artisan work. In other words, they are mostly part-time artisans. It may be mentioned that artisans in the Dahod villages have always had some land of their own. The blacksmiths were brought from other parts of the district and were given some land when settled in the village. The carpenters were full-time cultivators before they started carpentry work.

Also important is the fact that the patron-client relationship between cultivators and artisans has broken down in most cases. A shift from kind to cash transactions is associated with the disintegration of these traditional relationships. Earlier, artisans were given grain every year by the cultivators; mainly for repair work but in some cases even for fabricating implements. This customary share was given to the artisans even when they did not provide any services to a specific cultivator. This system, known as *grahki* is the Kalol villages and *hukdi* in the Dahod villages, is now nearly extinct. Only one artisan each in Gangardi and Ghoda, and the leather worker in Dadur, reported that they still have this arrangement with a few cultivators. The discontinuation of the *grahki* and *hukdi* practices has led to a decline in the assured income of the artisans and added to the instability of the market for their skills.

The other noteworthy feature is the growing competition in the agricultural implements market. The Gangardi blacksmiths, for example, are facing competition from migrant blacksmiths who come from neighbouring villages every day and work in rented sheds in the village. The weekly village fairs (*hat*) in Garbada provide an opportunity to the artisans of the region to come and sell their implements and adds to this growing competition. Smaller agricultural implements such as sickles, spades and axes are sold here. Significantly, no artisan from Gangardi sells his products in this *hat*. The expansion of fabrication facilities by the Garbada artisan has also adversely affected the Gangardi artisans. The Kalol village artisans on the other hand

Table 16: Agricultural implements/parts marketed by agents in the study region

| | Kisan Fertilizers | | Gangardi Co-operative | |
Implement	Number sold in 1981-2	Brand name/ source	Number sold in 1981-2	Brand name/ source
1	2	3	4	5
1. Hand hoes	250	Tata[a]	—	
2. Pickaxes	200	Tata/Jaihinda[a]	—	
3. Digging hoes	200	Local	—	
4. Rakes (I)	60	Local	—	
5. Winnowing fan	25	Varun[b]	—	
6. Dusters	27	KR-2 and KR-3	5	Saathi Gujkomasol Guj-S.D.
7. Sprayers	50	SR-4 and	5	—
8. Cotton stalk puller	25	SRP-4	—	—
9. Baroda hoe	7	Mamta[b]	5	Mamta[b]
10. Blade hoe/harrow (I)	5	Mamta[b]	6	Mamta[b]
11. Iron plough	—	Mamta[b]	—[f]	—
12. Blades (*pas*)[c]	200	Jaihind/Kisan[b]	—	—
13. Iron plough share	—	Local	—	Local
14. *Dadha*[d]	200	—	—	—
15. Hand pump	—	Local	—	Bardol[e]
16. PVC pipe	—	—	—	Gujarat Agro-Industries
17. Oil Engine	—	—	1	Ajit, Kirloskar

Notes:

a Procured from dealers in Godhra and Nadiad.
b Procured from Mamta Iron Works, Billimora.
c Blades for blade harrow and hoe.
d An iron tube fixed in the head piece of the blade harrow/hoe
e and in which the blade is fixed.
e Procured from Yantra Vidyalaya, Bardoli.
f The Gangardi Co-operative has stopped selling iron ploughs made by Mamta Iron Works, Billimora because of loack of demand.

52

face competition from large scale manufacturers (e.g., Tata) of some traditional hand tools like spades. To the extent bullock-drawn iron implements have become popular, the market of village artisans has also contracted. The Dadur leatherworker, for example, now attaches a leather pipe (*sunth*) to the iron *moht* for those who use iron *moht*. Earlier he used to get more orders for leather *mohts*. Similarly some other new implements are repaired by the local artisans.

Changes in artisan households in response to changing market conditions

Changing market conditions have resulted in a change in the production functions of the artisans facing disintegration. Of the two functions, the production and the repair and maintenance of agricultural implements, the relative importance of the latter function seems to be increasing while the former is gradually on the decline. The other response to the changing market situation faced by the artisan families has been an occupational shift among these households. As reported earlier, the younger workers of these households have moved away from their traditional occupations and have taken up salaried jobs mainly as turners and welders. Of the 27 sons of the present heads of artisan households (blacksmiths and carpenters), only 11 (41 per cent) have taken up their fathers' vocation. If we exclude two artisan households which are now based in Kalol town, this proportion declines to 26 per cent (five out of 19).

Not all artisans, however, face disintegration. Some artisans acquire new fabrication facilities and start operating on a somewhat larger scale, as in the case of Kalol town and Garbada artisans. In some cases they even become ancillaries to larger scale units.[18] Significantly, these households also do not experience the occupational shift we observed among other artisan households. But in the process of becoming a larger unit, these households cease to be self-employed enterprises and start hiring labour. It is very difficult to identify the characteristics that distinguish the artisans who have prospered over time from those who face extinction. However, all the relatively prosperous artisans are based in a large centrally located village or town where the size of the market permits them to fabricate implements on a somewhat higher scale. The individuals in the two artisan families of Kalol who had actually initiated the use of mechanized fabrication facilities had experience of working in large factories. The emergence of mechanized units mitigates, to a certain extent, the supply bottlenecks of fabrication and repair of agricultural implements. However, as we shall see later, these changes in the organization of artisan production do not remove all the supply constraints, which still inhibit the diffusion process. Anticipating the discussion in a later chapter, we may also note another kind of artisan response to the changing market conditions which helps him to remain competitive in the market, at least for a short period. This response relates to the modifications the artisans make in the old and new agricultural implements, in response to the changing cultivation practices and to make the implement more suitable to the region's agro-climatic conditions. Such adaptations and modifications have a positive influence on the diffusion process.

To conclude, three major points stand out from the discussion in this chapter:

○ Villages with higher agricultural productivity and levels of commercialization use more varied and modern agricultural implements.

○ An analysis of the ownership patterns and the sharing, borrowing and

hiring arrangements, suggests that three types of new implements are mainly used in the selected villages. These include manually operated modern implements like sprayers, dusters and winnowing fans, a few bullock-drawn iron implements and some power-driven implements. All these implements are mainly supplied by sources which are outside the traditional network of local artisans.

○ The suppliers of agricultural implements, particularly the village artisans, face an unstable and contracting market situation, which has led to occupational shifts away from their traditional occupations.

Notes

1. Some of the medium and large farmers in Ghoda supply sickles to the labourers when they employ them for harvesting their crops. They do so as they do not like the quality of sickles owned by the labourers. That is why one finds a high number of sickles owned per household. It is interesting to note in this context that the number of sickles owned per household is also affected by the number of working men, women, and children in the family; the size of the sickle varies according to the age of the worker.
2. Blanks against seeding funnels (*nalichala* and *orni* in Table 10 for Sureli) are misleading. Sureli was the first village we surveyed. The cultivators normally consider *nalichalas* and *ornis* to be parts of seed-drills and therefore did not provide information for them separately. Only later we realized that these 'parts' are normally bought separately and not necessarily along with the seed-drills. We corrected this mistake in the remaining three villages. In general one can assume, however, that a cultivator who has a two- or three-tined seed-drill will always have *nalichala* and *orni* and in the case of Sureli, especially *nalichala*.
3. There is no systematic relationship between size of holding on the one hand and the sources of agricultural implements on the other. To save space, the relevant data on this point are not reported here.
4. The 'purchase price' estimates were more severely affected by these reporting problems than the estimates of 'current prices'. Thus, although we did collect data on 'purchase prices' only the estimates of 'current prices' are reported and analysed here. The respondents were asked to report the current price of agricultural implements of the same quality and specification which they themselves own.
5. It is also likely that the idea about the current prices is affected by when the implement was bought, its price and how often its various parts were replaced. This bias may be particularly important in the case of those cultivators whose market contacts are very limited.
6. Differentials in reported current prices across cultivator categories were not very marked and no systematic relationship between size of holding and reported prices could be discerned. The tables bearing on this point, therefore, are not presented here. They are available on request. In any case, the number of observations in each sub-category was not large enough to draw any firm conclusions. A couple of interesting observations may, however, be noted: In general, for bullock-drawn implements there were more cases of a marginal farmer reporting

a relatively lower price than a small, medium or large cultivator; and for almost all the manually operated traditional implements, prices reported by small and marginal farmers were either comparable to those reported by medium and large farmers or higher. One speculative explanation for this can be that the small and marginal farmers use these tools more intensely than other farmers as they carry these implements with them when they go out to work as manual labourers. Consequently, these farmers may prefer better quality, more durable and, therefore, more expensive handtools than their wealthier counterparts.

7. For this table only, those responses were considered which also provided the purchase prices; the age figures without the corresponding purchase price were dropped. The age distribution of implements by size group of holdings is not presented here but is available on request.

8. Only one case of hiring a duster for Rs1 a day was reported.

9. There was one case of an electric motor being shared by two marginal farmers (brothers) in Sureli.

10. Marginal farmers mainly borrow from other marginal farmers, particularly traditional handtools. The bullock-drawn implements are sometimes borrowed from small cultivators as well (see Appendix 2). Interestingly, a barber/cultivator in Gangardi borrowed bullocks from his customers.

11. The Gangardi farmer also provided cigarettes and tea.

12. While the sample cultivators of Ghoda usually paid for hiring the services of a tractor for ploughing on a per acre basis, in Gangardi and Dadur they paid on an hourly basis. We were told that in an hour a tractor can plough slightly more than one acre.

13. Hiring of the services of electric motors for irrigation was more common among marginal and small farmers in Sureli.

14. The WITCO match factory is one of their regular clients.

15. The Dadur carpenter seems to have overestimated his sales. He said that he mainly sells in Dadur village which is quite small. The reasons for this over-estimation are unknown. Perhaps he felt we were assessing his eligibility for loans.

16. At times, even coal is provided by the customers to the Ghoda artisans.

17. This is very important for them because they normally face difficulties in procuring these raw materials, either because of resource constraints or because of supply bottlenecks. The Gangardi blacksmith reported that at times they cannot even get coal.

18. As already noted, the Kalol town fabricators had mechanized fabrication facilities and were undertaking piece-work for other production units.

Chapter 4
Diffusion of Agromechanical Technology: Correlates and Constraints

Some changes in the use of agricultural implements were identified in the last chapter. It was noted that villages with higher agricultural productivity use more sophisticated and improved versions of agricultural implements. Many studies have highlighted that a widespread use of HYV technology is associated with agricultural development. The regional and size-class biases in the adoption of biochemical technology are also well documented (For a review, see Dasgupta, 1977). Except with respect to tractorization, these biases in the diffusion of agro-mechanical technology have not generally been analysed. This chapter attempts to document some of these differentials while analysing sources of diffusion of agro-mechanical technology and identifying constraints to such diffusion.

Changes in techniques and processes of agricultural production may provide an impetus for improvements and changes to the agricultural implements, and vice versa. The nature and extent of irrigation, cropping patterns or sowing techniques may, for example, lead to changes in agricultural implements. Use of HYV seeds, chemical fertilizers and pesticides may also affect the use of agricultural implements. It is possible, however, for the agricultural implements to change without any corresponding change in the agricultural processes. New implements may be adopted on the basis of efficiency and cost considerations. But calculations of efficiency and cost are influenced by the extent of commercialization, size of the output or operation, available supplies of labour, etc. Lower requirements of labour may not be an important consideration for farmers and/or regions with labour supply surplus to demand. Information flows and development of views and beliefs emanating from the interaction among cultivators, artisans, extension personnel and agents operating in the region may also facilitate the use of new implements.

It is very difficult to separate out the influence of various forces enumerated above. However, to facilitate analysis and to bring out crucial issues more clearly, we shall discuss the role played by information flows and interaction among cultivators, artisans and extension personnel in the process of technology diffusion in the next chapter. This chapter discusses the other dimensions of the diffusion process. The first part relates changes in the nature of agricultural production in the study villages with changes in agricultural implements through specific cases as reported by the farmers, and the second enumerates the reasons for non-use of new implements.

Biochemical practices and the diffusion of agro-mechanical technology

An attempt was made in the survey to relate the changes in use of agro-mechanical technology to changes in biochemical and other agricultural practices. After identifying changes in the various agricultural practices relating to seed-fertilizer technology, respondents were asked whether any of these changes had been associated with a change in the use of agricultural implements. The results of this exercise are summarized in Table 17. Changes in agricultural implements associated with changes in the cropping pattern, sowing practices, use of HYV seeds and chemical fertilizers, etc., are reported separately. Along with the number of farmers who introduced the change, their distribution by size group of land holdings is also reported in Table 17. The relatively small number of responses does not permit any detailed statistical analysis. In most cases, it also does not permit any significant conclusions regarding the differentials across size groups of land holdings. However, wherever possible we have identified these differentials and drawn tentative conclusions which can probably be used as hypotheses in later studies. The cases reported in Table 17 underline the links between biochemical practices and the diffusion of agro-mechanical technology. The data effectively highlights various dimensions of the relationship between the adoption of biochemical inputs and changes in agro-mechanical technology at the farm level.

If there is any complementarity between the use of bio-chemical inputs and the use of agricultural implements, the cultivators can respond to a given change in the biochemical practices by choosing from the four options (I-IV) or by making no change at all (i.e., non-adoption). This typology of changes in agricultural implements reflects the differences in cultivator responses and is, therefore, useful.

Without going into the details of every change reported in Table 17, we discuss below some typical cases to highlight the main features of the changes introduced by the respondent cultivators. The table provides many instances of old implements (with or without modification) being put to use more intensively. In some cases, the cultivator knows about the implement but starts using it when he grows a new crop or adopts a new process. Others use an implement more intensively for new crops and processes, or put the whole implement, or parts thereof, to new uses, e.g., using seed-drills for making furrows, for dibbling seeds and for applying chemical fertilizers. Ploughs are used to make furrows in Dahod villages. These changes have, on the one hand, restricted the use of two- and three-tine seed-drills to making furrows

Table 17: Biochemical practices and the diffusion of agro-mechanical technology

Nature of change in biochemical practices	Associated changes in agricultural implements	Village (code)[1]	Size class[2] and number[3] of cultivators introducing change	Type of diffusion[4]
1	2	3	4	5
New crops				
1. Cultivation of groundnuts	Use of blade harrow for land preparation	2	Small (1) IV	
2. Groundnut	Use of *karban* (a bigger and heavier version of blade harrow) for harvesting	2	Small (1)	IV
3. Cotton	Removal of the middle tine of the old three-tine seed-drill and its use for making furrows to dibble cotton seeds	2	Small (1)	III, IV
4. Cotton	Use of cotton stalk puller	1 2	Large (1) Medium (1)	I
Sowing practices				
5. Shift from row sowing or broadcasting to transplanting of paddy	Use of bullock-drawn puddler	1 2	All cusltivators transplanting paddy (18)[5] Six out of seven transplanting paddy[5]	I
6. From broadcasting transplanting to row sowing of paddy	Use of three-tine seed-drill	3	Small (2)	IV
7. Broadcasting instead of row sowing of *bavto* (finger millet)	Use of planks for the crop, entailing deeper ploughing	3	Small (1)	IV

8.	Dibbling of naize seeds	Use of two/three-tine seed-drills for making furrows	1,2	All HYV maize growers	IV
9.	Increase in irrigation	Use of bullock-drawn bundformers instead of rakes and spades for making bunds	1	Most of the cultivators	I
10.	Use of chemical fertilizer	Use of whole seed-drills for putting di-ammonium phosphate and seeding funnels for urea[6]	1 2 3 4	All users (100%) Most (75%) All (100%) Majority (63%)	IV
11.	Use of pesticide	Use of dusters and/or sprayers	1, 2, 3 4	All users Some users (43%)	I

HYV Seeds

12.	HYV wheat	Use of new three-tine seed-drill, the distance between two tines being 9" and special blade hoe for this distance	2	Large (1)	I, III
13	Use of HYV seeds for pearl millet and maize	i) Use of olf three-tine seed-drill with middle tine removed	1 2	Marginal (1) Medium (3) Marginal (1) Small (1)	III
		ii) Use of new three-tine seed-drill with middle tine removed	1	Medium (3)	I, III
		iii) Shift from three-tine seed-drill to two-tine seed-drill	1	Small (4)[7] Medium (3)	
		a) Same seed-drill for both crops	2	Marginal (1)[8] Small (1) Medium (1)	I, IV
		b) Separate two-tine seed-drills for maize and pearl millet along with new blade hoes	2	Large (1)	

Table 17 (continued)

Notes:
1. Village code: Ghoda - 1; Sureli - 2; Gangardi - 3; Dadur - 4;
2. Size class of cultivator: Marginal (0-1 hectare); small (1-3 hectares); medium (3-7.5 hectares); and large (more than 7.5 hectares)
3. Except with respect to the use of chemical fertilizers (Case 10) figures in parentheses in column four show the number of cultivators who reported the change in column 2
4. Type of diffusion: I - Adoption of a new agricultural implement;
   II - Use of a modified version of the new agricultural implement;
   III - Use of a modified version of an old agricultural implement; and
   IV - Use of an old agricultural implement for a new purpose.
5. All paddy growers transplant paddy in village 1. In village 3, while all medium and large farmers growing paddy transplant it, only 31 per cent of small and marginal paddy growers do so.
6. In Kalol villages seed-drills are used to make furrows while the same operation is performed with ploughs in Dahod villages. Whole seed-drills are not used in Dahod villages for the application of chemical fertilizers. Only *orni* is used.
7. Two of these cultivators use the two-tine seed-drill only for millet; for maize, they use the old three-tine seed-drill with the middle tine removed.
8. This cultivator first used his old three-tine seed-drill with the middle tine blocked. This increased the distance between rows from 16" to 32". Later, he acquired a two-tine seed-drill with a distance of 28" between tines.

Table 17 (continued)

instead of actual sowing in the Kalol villages and, on the other, have increased the variety of uses of the implement to include the application of chemical fertilizers, etc. Somewhat similar is the use of ploughs and single-hole seeding-funnels in the Dahod villages.

Irrigation and changes in agricultural implements: We have already noted in the last chapter that the use of new irrigation equipment such as oil engines and electric motors is on the increase, while the use of *rahats* and *mohts* is on the decline. This process is more marked in Kalol villages (especially Ghoda) than in Dahod villages where *mohts* are still used on a fairly large scale. There has been a shift from leather *mohts* to iron *mohts*, apart from the shift from *mohts* to oil engines and electric motors.

In a few cases, increases in irrigation have also been associated with the use of bundformers in Ghoda (village code 1 in Table 17). We have already seen that bullock-drawn bundformers are quite popular in this village. Bundformers are more efficient than rakes and spades, and save a lot of time and labour.[1] Availability of water has also led to a shift from broadcasting to transplanting paddy. This, in turn, has led to use of puddlers substituting for ploughs and seed-drills. Fall in water supply or uncertainties regarding water availability induce shifts in the opposite direction, that is, from transplanting to row sowing, which have led to increased use of seed-drills. Thus, some changes lead to a decline in the use of a specific implement, while some others enhance its use. At any one time, the intensity of use of an implement is determined by the balance of these countervailing forces.

Use of HYV seeds and changes in agricultural implements: From the point of view of diversity of the diffusion process, the adoption of HYV seeds for maize and pearl millet (*bajri*)[2] and the related changes in the use of seed-drills and blade hoes, perhaps provide the most interesting case. The HYV varieties of maize and summer millet require greater distance between rows than the traditional varieties. The prescribed distance between rows of millet is 24" and for HYV maize it is 28". This has led to some changes in implements. These crops were earlier sown by a three-tine seed-drill when the distance between the two tines was normally 16".[3] Apart from dibbling[4] maize seeds, cultivators in Ghoda and Sureli (Villages 1 and 2 respectively in Table 17) have responded broadly in five different ways to this problem. [5]

O Some cultivators remove the middle tine of the old three-tine seed-drill for sowing maize and millet and thus increase the distance between rows from 16" to 32"

O Some use a new two-tine seed-drill with tines 24" or 28" apart. They acquire a two-tine seed-drill for either of these two sizes and use it for both crops.

O Some obtain a new three-tine seed-drill, with distance between the two extreme tines ranging from 24" to 28". After removing the middle tine, this seed-drill is used for both maize and millet.

○ Others make separate two-tine seed-drills of 24" and 28" for millet and maize respectively.

○ Finally there are cultivators who do not make any changes in their seed-drills.

It is significant that a majority of small and marginal farmers in Sureli have not made any changes in the implements and maintain the original distances. Only one large-scale farmer has opted for two separate seed-drills for maize and millet; others have restricted themselves to less costly options. The large cultivators have also purchased new blade toes with specifications to match the new seed-drills. In Ghoda, medium-sized farmers have made more changes than the small and marginal farmers, but in general the proportion of cultivators not making any changes is lower there than in Sureli.[6] Adjustments, other than blocking (or removing) the middle tine of a three tine seed-drill, were reported by some cultivators in Ghoda. Apart from blocking the middle tine, they use pegs of wood to reduce the distance between the two extreme tines after removing the middle tine. Similar adjustments were reported for wheat by two other medium-sized cultivators. In this case, however, they use all the three tines but put pegs in the two extreme tines to reduce the distance between tines. One of these cultivators also bought a new blade hoe appropriate for these reduced distances.[7]

Another cultivator (medium) who had purchased a new set of seed-drills was making some interesting experiments with the distance between rows. His seed-drills have two tines which are wider apart than is traditional, but less than prescribed. He claims to get both more crop and fodder output by these adjustments and by applying a fertilizer dose slightly larger than prescribed.

Use of pesticides and changes in agricultural implements: The use of plant-protection equipment is associated with the use of pesticides. While cultivators in Sureli and Ghoda mainly use liquid pesticides, those in Gangardi and Dadur use both liquid and powder pesticides. Consequently, cultivators in the Dahod villages use both dusters and sprayers, while those in the Kalol villages mainly use sprayers.[8] Unlike in the other three villages, not all the cultivators who use pesticides in Dadur (the most backward village) use plant protection equipment. All the marginal farmers and 50 per cent of the small-scale cultivators broadcast it manually. Some cultivators have discontinued the use of pesticides. Crops treated with pesticides are considered to be unsuitable for family consumption by a few farmers. Others feel that if the pesticide is not washed off the plant by rain or irrigation water, neither the plant nor the weeds can be used as fodder. The cattle do not consume this fodder and, when they do, it reportedly affects their health.

The discussion so far has identified various types of changes in agricultural implements associated with changes in various cultivation practices. The relatively developed villages (Kalol villages) have experienced more changes in cultivation practices and implements than the less developed ones (Dahod villages). Moreover, the observed changes in agricultural implements are not uniform across cultivator categories and not all cultivators within the same category have responded to changes in agricultural practices in an identical manner. In fact, the biases of bio-chemical technology adoption seem to get accentuated and perpetuated through differentials in agro-chemical technology adoption. There is, therefore, no direct correspondence between the diffusion of bio-chemical and agro-mechanical technologies. As far as possible, the small-scale and marginal cultivators try to use the old implements (with or without modifications) to perform new tasks. In such cases, as older equipment is modified to meet new demands, the diffusion of new implements to perform new tasks is bound to be affected. For example, if the old seed-drill or its funnels are used to apply chemical fertilizer, new implements such as the seed fertilizer drill may not be immediately adopted unless it has considerable additional advantages. The use of traditional seed-drills for applying chemical fertilizers involves more intensive use of human and bullock labour. So long as family and bullock labour have low opportunity costs, this multiple use of the traditional seed-drill may continue to be the first preference of cultivators. In what follows we discuss some more factors which explain the differentials in agro-mechanical technology diffusion.

Inappropriateness of agricultural implements

The observed association between the conditions of peasant production and the use of various agricultural implements brings out the importance of demand factors in the process of agro-mechanical technology diffusion. Analysis of the reasons reported by the cultivators for not using specific sets of agricultural implements may help to explain the different responses of cultivators on the use of agro-mechanical technology under similar conditions of production.

The survey sought to collect detailed information on reasons for not using specific agricultural implements. The reasons were to be enumerated separately for each implement which the sample cultivators did not use but which was available in the region. However, at the time of the survey, the discussion always veered towards general issues. Consequently, a uniform pattern could not be discerned in the cultivators' responses. Therefore, we cannot organize the responses of the cultivators in the form of tables. The responses reflected various dimensions of the inappropriateness of the available agricultural implements. In what follows we highlight these dimensions.

General issues: Agro-mechanical technology does not affect yield directly. Yields associated with the use of different agricultural implements are variable and depend on rainfall, fertility of the soil, use of other agricultural inputs, timing of agricultural operations, and so on. Under such circumstances, any calculation of the relative efficiency of various agricultural implements is complex and can only be approximate. However, since the purchase of agricultural implements involves fixed investment (in some cases large), one would expect the cost of the agricultural implement to be a major factor in determining its diffusion. Indeed, cultivators complained that the new implements, even bullock-drawn iron implements, are costly not only in terms of initial investment but also in terms of the costs of repairs and replacement of parts. The responses indicated the cultivators' unwillingness to become dependent on the market, which is outside the traditional artisan network. They somehow feel that the dependence on the 'outside' market will mean less control over their farm production. Each cultivator grows those crops and uses that set of implements which after (more or less continuous) conscious deliberation, appears to be best suited to his needs and economic situation. Reluctance to enter into market dealings and cash transaction, on a larger scale, was most marked in the case of cultivators of Dahod villages. Noteworthy in this context is the fact that none of the bullock-drawn iron implements or other improved implements are fabricated within the surveyed villages. These implements cannot even be repaired within the village.[9] Non-availability of fabrication and repair services near the village adds to the cost factor. Costs of replacing parts and repairs are higher when such facilities are not easily available as more time and money are required to get these services. Delay in repairs proves to be very costly in cases where particular operations have to be completed relatively promptly because of factors beyond the control of the farmer. The cultivators specifically said that the use of machines makes them overdependent on erratic power supplies and specialized repair facilities which are expensive and not easily available. Those who hire these implements also encounter uncertainty about their availability at the desired time. Except perhaps for tractors, very few machines can be used throughout the year and remain idle most of the time. Overall, high costs and uncertainties regarding the supplies of fabrication and repair facilities constrain the use of new agricultural equipment. Therefore, to the extent that the use of new implements involves high initial investments and dependence on outside market forces, cultivators are reluctant to adopt and use them.

Since the purchase and maintenance of various agricultural implements entails fixed and recurring expenditure, the farmers' capacity to spend these amounts is obviously important in determining the pace of their adoption. Land being the most important asset of the farmer, the size of landholding is a summary index of a large number of potentially important factors such as access to scarce agricultural inputs including credit, capacity to bear risks, access to information, etc. Thus all things being equal, large-scale farmers can be expected to adopt new agricultural equipment first. Our data does not permit any detailed analysis of the adoption process by size group of holdings. However, changes made in various agricultural implements following the introduction of HYV maize and millet very clearly suggest that the cultivators with smaller holdings opt for less costly options.

We noted in the earlier section that not all cultivators in Ghoda and Sureli changed the distance between rows for HYV seeds of maize and millets and even those who did, did not always conform to the prescribed distance; they resorted to partial adjustments such as removal of the middle tine of the three-tined seed drill, etc. Only the large cultivators bought new seed-drills for HYV cultivation. A change in distance between rows may mean more than what is evident on the surface. An increase in distance between rows, some cultivators argued, reduces the number of rows in a plot, which a small land holder cannot afford. In addition, if he adopts inter-cultivation practices, the fodder production in the farm may also be reduced. It was emphasized that the increase in the distance between rows does not lead to such an increase in crop yields that it can no more than offset the loss incurred due to fewer rows per unit of land and the decline in fodder availability. Cultivators argued that the distance between rows maintained by them optimizes food and fodder production and reduces their dependence on the market for either food or fodder.

In this context another interesting dimension needs further exploration: the complementarity between various implements or new specifications of implements can lead to more than one change in the use of implements. As already noted, few large farmers have also exchanged blade hoes for seed-drills. Some cultivators argued that a seed-drill with new specifications may require a new blade hoe, a new yoke and at times even a new blade harrow. Changing this whole set is very costly and, therefore, sometimes even when necessary, no changes are made in the specification of any of them. But such changes are necessary only when major changes are made in the specification of an implement. Some cultivators reported that although the size of their new seed-drill was decided by the size of the blade hoe and yoke that they already had, minor changes and adjustments are always possible. Apart from this problem of complementarity, the specifications of implements being used by other cultivators in the village restrict the extent to which a cultivator can change his own implement. The respondents argued that if one used implements with specifications quite different from those which others use, one cannot borrow or hire others' implements when needed urgently.

In farms which are primarily family-labour based, saving of labour is not a major objective. In the absence of alternative employment opportunities, cultivators do not impute any cost to family labour. This was particularly the case in Dahod village where 'income' normally meant only the cash which they get by selling their farm output or through wage labour. Thus considerations of saving family labour or efficiency are not very important for them. Processing equipment such as winnowing fans, threshers, and new bullock-drawn implements like bundformers, which are normally used for their efficiency, [10] are not very attractive to small and marginal farmers, whose output volume is too small to justify the cost. This is especially true for Dahod villages which have only one crop per year.[11] Similarly, for large cultivators the use of implements such as the groundnut opener is economically unattractive when cheap child labour is available.

Poor cultivators normally look for cheap multi-purpose implements.[12] Thus the chaff-cutter (*sudo*), a specialized implement for cutting fodder, is not used by the majority of cultivators because they do not have many animals and also because an axe, which is a multi-purpose implement, can do the job. Cultivators in Dahod villages also use sickles to cut fodder. In quite a few cases, the cultivators argued that the iron bullock-drawn implements are not perfect substitutes for the old wooden ones, since they are not as versatile.[13] In such circumstances, cultivators are forced to retain the old implements even when they buy new ones. Maintaining two sets of implements is very costly for cultivators owning a small landholding.[14]

Apart from the above-noted considerations the other set of factors leading to the non-use of new implements relate broadly to the operational inappropriateness of these implements, given the agro-climatic conditions of the region. Some parts of Kalol villages have heavy black soil which sticks to the bullock-drawn iron implements, making them heavier. These implements also go very deep (particularly during *Kharif* season when the land is soft due to rain), which is a tremendous strain on the bullocks. The operator also finds it very difficult to hold these implements, which are heavy to lift.[15] For Dahod villages the problem is somewhat different as the land there is stony and undulating. Ploughs, in general, break very often on this terrain. The use of iron ploughs, which are heavier and less manageable, is out of the question here. Even blade harrows and wooden seed-drills are normally not used because of the terrain.[16] The unevenness of the land adversely affects the uniformity in depth of the seeds sown. Also, when a plough is obstructed by a stone and the bullocks pull it, the plough may break if it is wooden; on the other hand, the bullocks may get hurt if the plough is made of iron.

The size of the bullocks adds another dimension to these operational problems. While the type of land constituted the major cause in Dahod villages, bullock size was the common cause mentioned in Kalol villages for not using bullock-drawn iron implements. This was despite the fact that bullocks in Dahod villages are smaller than bullocks in Kalol villages. The cultivators felt that the size of the bullocks normally used in the region is such that the draft power available will not be adequate to pull the heavy iron implements. The fact that some cultivators do use iron implements is explained by the prosperity of the users who are capable of providing more feed to their bullocks which a small cultivator cannot afford. It was argued that even if the bullocks are able to pull the iron implements, it will mean a reduction in their life-span and they may get hurt.

The type of land can influence the use of even those implements which are not heavy or bullock-drawn. In Dahod villages, non-serrated sickles are used, while the Kalol cultivators use serrated ones. In fact, no blacksmith in Dahod villages knows how to make a serrated sickle. It was argued that a serrated sickle is inappropriate for the stony terrain as the teeth will break often and the sickle will need serration at frequent intervals. While non-serrated ones can be sharpened easily with the help of a stone, serrated ones require the services of a blacksmith.

Problems specific to particular implements: The iron ploughs are considered more efficient than wooden ploughs of equivalent draft. To get the same degree of tilth that is obtained with a single ploughing with an iron plough, one would have to work the wooden plough four times at least and this requires three times more time and energy (India, NCA, 1976, pp.295-6). The majority of cultivators in the sample villages, however, argue that iron ploughs do not invert the soil, the way traditional ones do. At best, they have the same efficiency as the wooden ones but are much more costly. Similarly, it has been argued that tractor tillage is not deep and good enough and the cultivators have to use wooden ploughs after tractor ploughing which turns out to be very costly.[17]

For iron-blade harrows and hoes, similar problems were also reported: they do not invert the soil and consequently weeds are not destroyed. The iron-blade harrow also leaves clods of earth which have to be broken later.[18] For iron seed-drills, the normal complaint was that the nuts and bolts of these implements loosen and consequently the distance between rows ceases to be uniform.[19] By tradition, Sureli cultivators continue to use a very large, somewhat modified version of a blade harrow (*karban*) instead of shifting totally to common blade harrow (*karab*) used by Ghoda cultivators. We were told that the use of the *karban* is a *baria* tradition, a caste prevalent in Sureli, while Ghoda is a 'village of Patels'. Thus even though the *karab* needs less draft and manpower compared to the *karban*, the latter is still used in Sureli.[20] In this context, it may be mentioned that the Ghoda blade harrows are bigger and heavier than the ones used in Sureli. It is possible that the agro-climatic conditions and the prevailing cultivation practices may require two different-sized blade harrows in Sureli while one, a somewhat heavier version, may suffice in Ghoda. The respondents did not explicitly confirm this conjecture.

The groundnut decorticator is also not very efficient; it breaks the kernel tip and at times removes the seed cover as well. If these kernels are used as seeds, plant growth is impaired and the crop is more susceptible to pests. Complaints of a similar type were reported for a maize sheller. It is inefficient in the sense that not all the grain (seeds) is shelled and some seeds remain on the cob. It is cumbersome because only one maize cob can be put into the machine at one time. A stick, used to beat the maize cobs, is more efficient, the cultivators argued. Besides, if the crop output is limited, women and children at home can always shell maize cobs manually. Problems with iron *mohts* were reported as well. A leather pipe has to be attached with iron nails to the iron *moht* and the nails either break or tear the leather pipe. A leather *moht* with leather stitches is therefore more appropriate for the type of wells which exist in Dahod villages.

Thus, the reasons for not using the new implements vary from cultivator to cultivator and from implement to implement. Apart from the problem of high costs and low levels of resource availability in the region, the discussion in this section has brought out three dimensions of the appropriateness of the available agro-mechanical technology:

O Inappropriate to the conditions of production based on the calculations of the farmer. For example, some implements were considered inappropriate because of the small size of the farming operation. Others were inappropriate for a farmer who was not market-oriented but making an attempt to be self-sufficient in food and fodder.

O Inappropriateness resulting from inadequate repair and fabrication facilities in the region.

O Inappropriateness with regard to existing agro-climatic conditions and factors of production such as use of bullocks and other implements.

Notes

1. According to a few estimates given by the sample cultivators, bunding of one *vigha* (0.57 acre) requires about 5 men for a day if they are using rakes and spades. The same work, one (or two) men can do with bullocks and bundformers in an hour. Another estimate is that 8 men are needed for bunding an acre in a day with rakes, while only two men are needed to finish the work in half a day with bullocks and a bundformer.
2. In the remainder of the text, referred to as millet only.
3. Two types of measurement units were used by the respondents in the field: inches and feet and *angad* and *haath*. Roughly 1 *angad* = 0.8 inches and 1 *haath* = 1.5 feet. We have used inches and feet in the text after converting the measurements reported in other units. The reported measurements vary a great deal.
4. Dibbling involves putting seeds manually after making furrows or holes with seed-drills or any other agricultural implement.
5. Two- and three-tine seed-drills are not in use in the other two less-developed villages (Gangardi and Dadur, Villages 3 & 4 in Table 17). Sowing is done with a plough and a bamboo funnel, and therefore the distance between rows can be changed at will.
6. Two-tine seed-drills were more common in Ghoda than in Sureli even before the advent of HYV seeds. Secondly, in Ghoda distances between the two extreme tines of the traditional three-tine seed drill and between the tines of a two-tine seed-drill were closer to the prescribed ones for HYV maize-millet cultivation than the Sureli seed-drills. Thus a shift to two-tine seed-drills for HYV millet/maize was perhaps easier.
7. This is a case similar to that of the large-scale farmer in Sureli.
8. Only the large-scale farmer in Sureli, and one small and one medium farmer in Ghoda reported the use of both sprayers and dusters.
9. The problem of inadequate repair facilities near the village has been identified as one of the major factors contributing to the non-use of new implements (India, NCA, 1976, Chapter 51).
10. Some estimates for saving of labour with bundformers are provided in an earlier section. For winnowing fans one estimate given was that in the traditional method one needs six people, while to operate a fan only three people are needed. Another cultivator contended that even two people can operate the fan.
11. The size of the operation/process has been shown to be an important factor in agro-mechanical technology adoption processes. (For a historical review, see Binswanger, 1984).
12. This to an extent, is true even of the larger farmers in the backward villages. A larger farmer

uses the sprayer to spray pesticides as well as urea in Gangardi.
13. That the traditional wooden ploughs are multi-purpose while the iron ones are not is an accepted fact (India, NCA, 1976, p.396).
14. In general, it is exceedingly difficult to ascertain whether the cost-benefit ratio of a specific agricultural implement is attractive enough for the cultivator. The issues discussed here add to the complexity of the problem.
15. This was the predominant opinion. There were, however, conflicting views. Some cultivators using these implements found them suitable for the soil while some others did not. We could not ascertain whether or not this divergence of views was due to intra-village differences in soil.
16. There is evidence to suggest that these implements cannot be used in stony soils. (see for details, India, NCA, 1976, p.397)
17. Some cultivators, however, contradicted this statement and argued that even if one has to use wooden ploughs after tractor ploughing, the land preparation is accomplished more quickly with a tractor. Bullock ploughing is faster and easier after tractor ploughing, as the bullocks need less rest.
18. One Ghoda cultivator suggested that the iron-blade harrow can be made more effective by maintaining the traditional shape. Thin iron sheets, instead of thick heavy ones, may be used to give it that shape to keep it light. In the traditional design, clods are crushed.
19. Another problem, associated with iron implements, reported to us was that they get hot and are difficult to handle.
20. A *karban* needs three people to operate it, while a *karab* requires only two.

Chapter 5
Links between Technology Generation and Use

Historically, many of the earliest improvements in farm tools were made by users themselves or by local artisans who responded to their needs. The pioneers of the Japanese machinery industry were the carpenters and blacksmiths of the farm villages who were involved as a matter of policy in the process of improvement, adoption and innovation of farm implements (Hemmi, 1982). It is important to identify the local nature of such improvements and innovations since agricultural practices differ across regions. These changes are made in response to problems faced by cultivators in specific agro-climatic conditions. In order to analyse the interaction between agents of technology generation and the various categories of users in the economic system, and the consequent impact on the improvement and innovation of farm tools, we have identified four types of economic agents: cultivators using the implements; various types of economic agents: cultivators using the implements; various types of artisan and fabricator; selling agents as non-producing suppliers; and extension personnel including *gram sevak* (village level workers).[1]

Information was collected regarding the kind of discussions that take place among the above-mentioned agents. An attempt was made to identify concrete cases of interaction which led to some changes in agricultural implements. To get such information through recall is difficult and it is possible that we may not have been able to capture all types of interactions and changes in implements. We first summarize the responses of cultivators, artisans and agents regarding the nature of interaction. The second part discusses concrete cases of improvements and innovations in farm tools, enumerated in our field survey.

Nature of interaction

In general, the discussions among these economic agents relate more to the seed-fertilizer technology than agro-mechanical technology. Experiences with HYV seeds, chemical fertilizers, pesticides etc., are discussed more than problems related to agricultural implements, both new and old. The cultivators' discussions with the extension personnel, for example, especially the *gram sevak*, mainly revolved around the possibilities of getting HYV seeds, pesticides and chemical fertilizers. Cultivators sought information about dosage and timing of applications of fertilizer and other related practices of seed fertilizer technology. That these inputs are not easily available to small and marginal farmers was clearly brought out by farmers' complaints. Some farmers have stopped using the new biochemical technology inputs because of supply problems. Such problems are more acute in the backward villages of Sureli, Gangardi and Dadur; the Ghoda cultivators are better off with respect to supply of these inputs. Similar problems exist with regard to the subsidy scheme through which extension personnel are supposed to popularize the use of improved agricultural implements. Information regarding subsidy schemes was not easily forthcoming, especially for small and marginal farmers. Cultivators complained that extension personnel come only to give 'lectures' and not to help. Cultivators in backward villages, particularly in Dahod villages, seem to have a typical response to the extension personnel's 'instruction', 'direction' and 'lecturing'. They just ignore them. This perpetuates two kinds of beliefs: the extension officers become convinced that the cultivators are crazy, backward and beyond redemption; the cultivators' cynicism regarding 'government help' is also confirmed or accentuated.

The cynicism and anger of small and marginal farmers seemed justified when we found some plant protection and other equipment, owned by the group Gram Panchayat in Gangardi lying in a *taluka* panchayat official's houses. These implements were meant for lending and hiring out to needy cultivators.

The interaction between cultivators and agents is also inadequate. Of the two agents operating in the region, the Gangardi Co-operative is more successful in terms of exposing cultivators to new implements and new agricultural practices. Normally no after-sale service is provided by these agents and no attempt is made by them to ascertain the problems faced by cultivators who use implements supplied by them. Even if some cultivators complain about their implements, it is rarely communicated to the fabricators. Thus, interaction between cultivators and agents and extension personnel is such that first-hand information does not flow to the less privileged population, while block-level officers, and even agents, prefer to keep contact only with the richer groups. Our field experience tends to substantiate Hale's (1973) conclusion that " . . . in the highly stratified and faction-ridden village society, any bias in access to first-hand information is perpetuated in further diffusion, with no extensive information crossing stratum and faction layers" (*Dasgupta*, 1977, p.242).

Interaction between cultivators forms the basis of opinions that farmers have about various agricultural practices and new agricultural implements. To a large extent, reasons reported for not using certain implements (discussed in Chapter 4) are reflections of these discussions. Cultivators' opinions are not, however, always formed merely on hearsay evidence. In a number of cases, cultivators related their own experiences with different agricultural practices and implements. Quite a few cultivators in Gangardi and Sureli, for example, had borrowed iron ploughs and blade harrows, used them and found them unsuitable for their type of cultivation. Some had even purchased these implements but did not use them any more. Information about such experiences is spread through discussions among cultivators and contributes a great deal to the formation of cultivators' opinion regarding various agricultural practices.

Another important source of information for cultivators, especially in Dahod villages, is labourers and cultivators who migrate to Kheda, Ahmedabad and Baroda districts for work. These workers are exposed to various kinds of new implements in these districts and their experiences spark off discussions among cultivators about these implements. Even a simple tool like a sickle differs across regions — it is serrated in parts of Kalol, Kheda and other districts but non-serrated in Dahod villages. The migrants use serrated sickles when they migrate out but non-serrated ones in their own fields. Their experience has shown them that serrated sickles are not appropriate for Dahod villages. Through discussions other cultivators are also convinced of the validity of migrants' opinion. The main themes of cultivators' discussions include crops to be grown, problems related with the use of chemical fertilizers and pesticides,[2] distance between rows, etc. With regard to new implements, apart from cost, the other concerns are: depth of sowing and ploughing, adequacy of bullocks to pull them, and availability of draft power. we have already seen that these concerns were reflected in reasons reported for not using specific implements (Chapter 4).

The interaction between the artisans and the cultivators is the most important as far as the improvements and changes in agricultural implements are concerned. Surprisingly few cultivators reported that they discuss new implements with the artisans. Instead, operational problems faced by the cultivators with different implements form the basis of discussion. Cultivators mainly come to the artisans with some specific problems, for adjustments in implements or for repairs. Their attempts to solve these problems lead the artisans either to improving the implements with some modifications or to some other type of innovation. The cultivators also communicate to artisans about the type of steel or wood to be used in their implements and the size and/or weight of the implement needed. The type of implement ordered is determined by factors such as the size of the customer's bullock as well as the age and sex of the person who will be using the implement. For example, we were told that even the size of the *moht* depends on the size of the bullock and sickle size depends on the age and sex of the user. In what follows we discuss some specific cases of those innovations and adaptations which emerged due to the interaction between artisans and cultivators and consequently contributed to the diffusion of agricultural implements.

Table 18: Artisan-cultivator interaction: some cases of changes by fabricators

Nature of change in agricultural implement	Problem with the earlier version	Whether the modification was tried yes/no	If yes, has it been accepted by cultivators[1]
Design changes			
1 Automatic seed-drill, which can control seed-rate. (Kalol town carpenter and blacksmith)	The new HYV seeds are costly and there is wastage of seeds as seed rate cannot be controlled with the old seed-drill	No	n/a
2 A straightened and heavy sickle (Gangardi blacksmiths)	The original sickle with a small radius could not cut sturdier crops and cutting of grass was difficult because of stony ground	Yes	Very popular can cut wood as well
3 New thresher design: the meter stops automatically if there is any obstruction (Kalol town blacksmith)	Any obstruction to the fan (stone etc.) could damage the motor	Yes	Extent not known
4 Use of four iron beams instead of one, to attach the winnowing fan to the stand. (Kalol town blacksmith)	The body of the original fan was unstable	Yes	Extent not known
5 Well pulley with ball bearings to reduce friction (Garbada blacksmith)	The improved iron *mohβ* was heavier than the traditional leather one. It was, therefore, difficult for the bullocks to pull it	No	n/a
6 Converted the winnowing fan with a belt pulley system (Kalol town blacksmith)	The use of the winnowing fan with a belt pulley system was very strenuous and required a lot of effort	—	—
a) With ball bearings in the pulley wheels		Yes	Extent not known
b) With a gear system		Yes	Extent not known

	c) With a cycle chain system		Yes	Extent not known
	d) With a motor		Yes	Not popular as it is costly

Substitution of raw material/parts

7	Iron made bullock drawn puddler: teeth welded to to the beam instead of using bolts to fix them: (Garbada blacksmith)	When bolts were used the teeth used to loosen and get caught in the soil	Yes	Not known
8	Substitution of the iron beam of an iron blade hoe/harrow by a wooden one (Kalol town blacksmith)	The blade hoes/harrow with iron beams were found to be heavy	Yes	Extent not known
9	Substitution of the wooden beam by an iron one in the traditional blade hoe/harrow (Kalol town carpenter)	Wooden beams used to break iften. Good wood is difficult to get. Iron blade harrow is costly. Some cultivators prefer the traditional design	Yes	Extent not known
10	Use of iron angles in the axle of the traditional bullock cart which reduces the cost (Kalol town carpenter)	High cost of the bullock cart. Increases in wood prices	Yes	Extent not known
11	Use of iron beams instead of wooden ones in the cart[4] (Kalol town carpenter)	Non-availability of the right type of wood	Yes	Extent not known
12	Use of bushes instead of ball bearings in the cart wheel (Kalol town carpenter)	For small bullocks, the cart is heavy and difficult to pull	No	n/a

Addition of new parts

13	Attach a curved iron rod to a bamboo stick and use it as a *kadial* (a type of fork/rake) (Gangardi blacksmiths)	The original bamboo *kadial* was not durable and bamboo is not easily available	Yes	Very popular

Table 18 (continued)

No.	Modification	Remarks	Seen	Popularity
14	Tines of Baroda hoe made adjustable (10", 12" and 14") by using bolts and iron pegs (Garbada blacksmith)	In the original design the distance between tines was fixed at 14" and the farmers wanted some flexibility	Yes	Not known
15	Use of some bolts to make adjustments in the angle of iron plough and blade harrow (Kalol town blacksmith)	The cultivators preferred flexibility so that they can change the angle according to size of the bullocks, crops grown and other agro-climatic conditions	Yes	Extent not known
16	Use of bolts to make the beam of the iron plough and blade harrow detachable so that for repairs the ploughshare can be carried separately (Kalol town blacksmith)	The farmers found it difficult to carry the whole iron implement	Yes	Extent not known
17	Fixing of thin iron sheets on the traditional wooden plough and blade harow to make it more durable without changing the original design or increasing the weight (Kalol town blacksmith)	The traditional wooden plough and harrows are less durable. Wood has become very costly. Farmers, however, prefer the traditional design to that of the new iron implements, besides finding the latter heavy	Yes	Extent not known

Notes:
1. Remarks in this column are based on our observations in the field. If we did not see the modified implement with any cultivator and only the artisan reported it, we have recorded it as 'not known'. If only one or two of the modified implements were seen, we use 'extent not known'. Only if more than five units of the modified implement are seen then we term it as 'popular' or 'very popular'. Sometimes the information supplied by the artisan was also used to guage the popularity of the modified implements.
2. Only a blueprint has been prepared.
3. The *moht* is a bag, traditionally made of leather and used for well irrigation. Bullocks pull it.
4. The carpenter who reported it was not very happy with this modification as the weight of the cart increased considerably.

Table 18 (continued)

Artisan/cultivator interaction

Table 18 provides details of some changes made by the local artisans to the agricultural implements. Problems faced by cultivators with specific implements, which provided the impetus for these changes, are also enumerated. The efforts of local artisans typically involved the following types of changes:[3]

O Changes in design
O Substitution of production material (or parts)
O Addition of new parts.

These changes are not mutually exclusive and as can be seen in Table 18, combinations of such changes do occur. Moreover, these changes are specific to regions, cultivators and agricultural implements. The addition of new parts and substitution of parts/material are two important options, apart from changing the design, utilized by the artisans to improve the efficiency and suitability of agricultural implements. That the scope for such changes is enormous has been brought out by the cases enumerated in Table 18.

Apart from solving operational problems, these modifications mainly increase the flexibility of the implement and make it more multipurpose. The durability of the implement is also improved, and at the same time the traditional design of the implement, which the cultivators prefer, is maintained.[4] This does not mean, however, that design changes are not introduced. Apart from other changes enumerated in Table 18, the idea of a ball bearing pulley is interesting and has been accepted elsewhere.[5] This pulley considerably reduces the effort required to pull water from wells. If this change is introduced and accepted, it will give a new lease of life to bullock-drawn irrigation equipment in the region, particularly iron *mohts* (water bags) which are modern and more durable versions of the tradition leather *mohts*. Consequently the adoption of oil engines and electric motors may be delayed.

To conclude, the discussion on agro-mechanical technology is mainly restricted to cultivators and artisans and neither the outside agents nor the extension personnel play any role in the adaptation and modification of agricultural implements. The interaction between cultivators and artisans at the local level does lead to some changes in the agricultural implements. These changes may be minor but they are relevant and important for the cultivators as they provide immediate solutions to their operational problems and thereby make possible more intensive and extensive use of implements.

Notes

1. Extension personnel are involved in the sale of HYV seeds, pesticides and chemical fertilizers which are at times sold through the Panchayat. They are also responsible for various subsidy schemes including those which are related with the sale of improved agricultural implements. It is noteworthy that a paddy transplanter, designed by a *gram sevak* in Orissa has been introduced in paddy areas recently (India, NCA, 1976, p.402).
2. There is a strong belief among cultivators in Dahod villages and Sureli that use of chemical fertilizers increases pest attacks. Pests become immune to different types of pesticides every year and unless new pesticides come into the market the cultivators are doomed. This is one

of the reasons why they do not use chemical fertilizers and pesticides.
3. Conceptually all these changes can be subsumed under 'design changes', the three-fold categorization is used mainly to facilitate presentation and to bring out the details of the modifications made.
4. Some carpenters argued for a selective substitution of wooden parts by iron ones in the traditional implements without changing their basic design. For example, according to them most parts of the traditional plough can be changed but the wooden ploughshare should be retained. Alternatively a wooden ploughshare may be fixed in an iron plough.
5. See Council for Advancement of Rural Technology (1985). The respondent artisan emphasized that it was his own idea and he wants to try it in his village. In spite of the fact that we later found the idea documented in the above-mentioned publication, we can give him the benefit of the doubt.

THE PROCESS OF AGRO-MECHANICAL TECHNOLOGY DIFFUSION

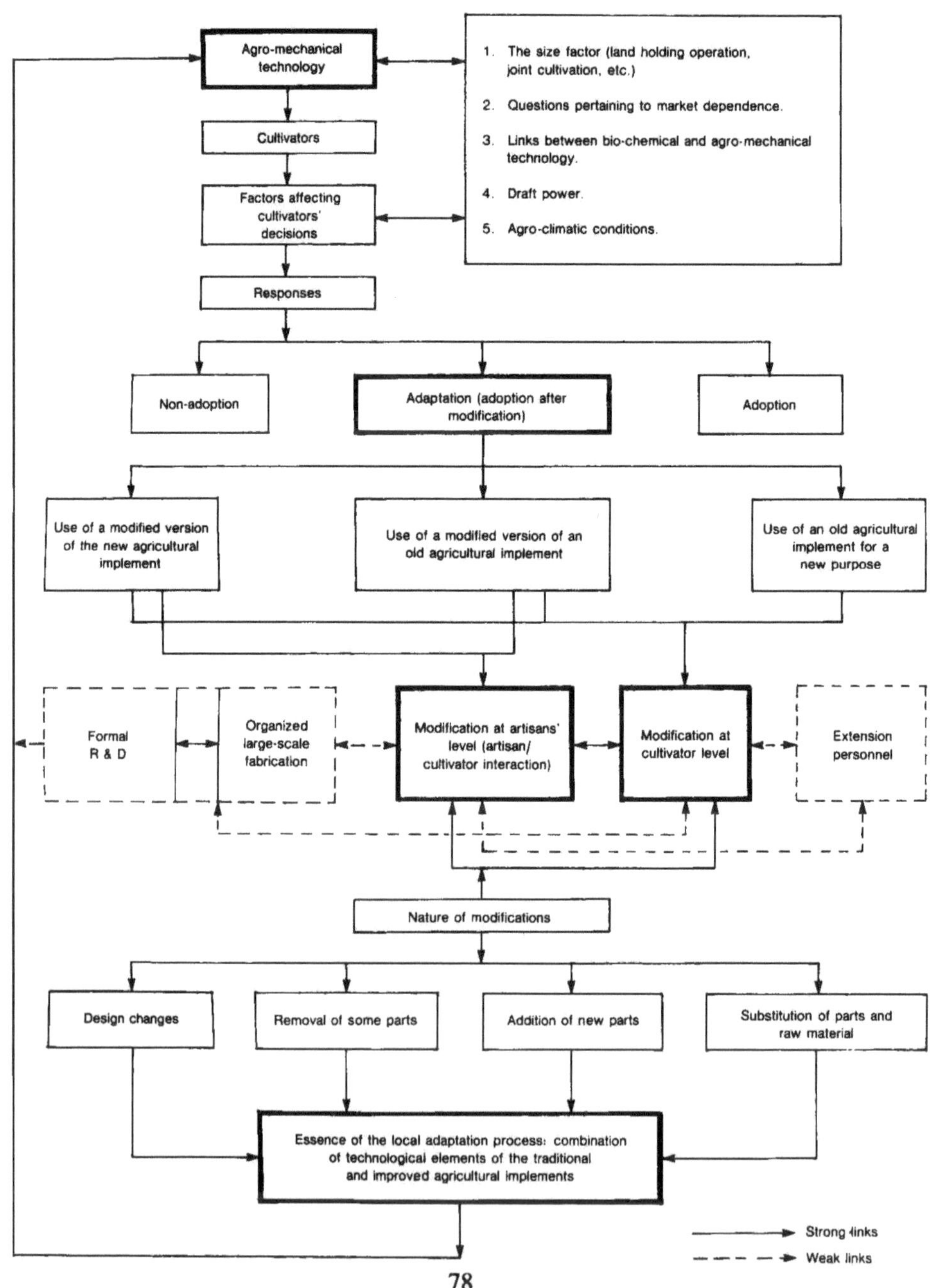

Chapter 6
Summary and Conclusions

The development and diffusion of agricultural technology, of which agro-mechanical technology is an important component, is a prerequisite for rural development. Historically, many of the earliest improvements in farm tools were made by users themselves or by local artisans who responded to the cultivator's needs. These changes were technically simple and involved only minor improvements in the available tool-making skills (Hemmi, 1982). The history of agricultural tools and machines also shows that even after the practical utility of an invention is finally approved it often takes a long time before it is generally used (Bath, 1960). The existing economic, social, infrastructural, agro-climatic and other conditions influence the pace at which a new implement gains general acceptance. Given these parameters, an attempt was made in this study to explore the conditions and relationships which characterize the process of general application or diffusion of agro-mechanical technology in a backward region.

An analysis of the process of technology diffusion may be divided into two components: analysis of the major factors which condition the responses of the technology users, and an analysis of the nature of the responses themselves. These two are interlinked. The results of our study on these two issues can be pieced together to develop a tentative outline of the agro-mechanical diffusion process and thereby generate testable hypotheses.

Factors affecting agro-mechanical technology diffusion

It is difficult to understand the logic of adopting new implements at farm level since they cannot be directly linked to increases in yields or cropping intensities. The process of agro-mechanical technology generation is complex: various interacting factors influence decisions about the use of an agricultural implement. In most cases the combination of influencing factors is specific to individual implements.

Size of holding: The size-class bias in favour of larger land-holding groups exists not only for agro-machinery but also for smaller, less costly implements. The family labour supply, relative to demand, and the opportunity cost of capital relative to that of labour, is higher for backward villages and smaller holdings compared to developed villages and larger holdings. To the extent, therefore, that the new implement is primarily a substitute for labour, it will not be very attractive to small cultivators and cultivators in backward villages. While in the backward villages, the system of joint cultivation increases the supply of family labour relative to demand, higher irrigation and labour intensive cropping pattern in the developed villages enhances the demand of labour with respect to supply (chapter 2).

Small-sized plots may discourage the use of seed-drills, bund-formers, etc., as the number of rows in the plot or the effective area available for tillage may be very small. We may recall in this context that the average size of the operational holding in the most developed village of our sample (Ghoda) was much higher than other villages (chapter 2). The economic advantage of introducing new equipment also depends on the size of the operation which, in general, is small for smaller holdings. This is an important variable affecting the use of post-harvest processing equipment. Hiring, borrowing and sharing arrangements take care of these size class biases only to a limited extent.

Questions pertaining to market dependence: The cultivators in our study villages (especially in the backward villages of Dahod), prefer to remain self-sufficient and resist entering into market dealings. This preference is reflected also in the cropping patterns of cultivators. Even medium and large farmers in the backward villages make an attempt to be self sufficient in food, fodder and seeds (Chapter 3). A desire to use weeds as fodder affects the extent of weed control activity[1] which in turn may affect the use of weeding and intercultivating implements. In cases where cultivators make their own implements, using the wood from their own farms, the degree of self sufficiency is even higher.

Some previous experiences of uncertainties with respect to the supply of HYV seeds and chemical fertilizers, and the non-availability or fabrication and repair facilities in the vicinity, probably reinforces the cultivators' fears about market dependence.

Links between bio-chemical and agro-mechanical technology: As mentioned earlier, it is extremely difficult to ascertain the economic viability of a technique. The problem is particularly complex in the case of agro-mechanical technology as it does not directly contribute to yields. However, cultivators do seek to economize on the use of scarce capital. Some important dimensions of the economics of agro-mechanical technology diffusion are brought out by its links with the use of bio-chemical technology. Our study observed a positive association between the use of seed-fertilizer-irrigation technology inputs and improved agricultural implements. The investment in machinery and implements may increase because of the need for weeding and intercultivating, etc., more intensively in HYV cultivation. In some cases changes in sowing techniques and cropping patterns necessitate the use of new implements and/or old implements with new specifications (see chapter 4 for various instances of this kind). Changes in agricultural practices may lead to not only one but a series of changes in the specifications of various implements in use, as well as the introduction of new implements.[2] The relationship between various implements in use is therefore a crucial component of the thinking of cultivators who are planning to use HYV technology and/or new implements since, in certain cases, a whole set of implements may have to be changed. To study the diffusion process of an agricultural implement in isolation may be analytically misleading.

Economy in the use of scarce capital is an important concern of farmers. This is reflected in the marked preference for multipurpose implements in the study villages. It is possible that the pay-offs from using specific implements for doing a particular operation for various categories of farms (or villages) differ according to differences in cropping pattern, irrigation, fertilizer use and other agricultural practices across these size categories (villages).[3] Given the links between the two types of technology, it is likely that implements which are developed to reinforce the effectiveness of seed-fertilizer technology are inappropriate and inaccessible to certain sections of rural population in the same way as the components of the seed-fertilizer technology. Moreover, the non-availability of HYV technology inputs may constrain the use of new agricultural implements and may further perpetuate size-class biases. The relationships between bio-chemical and agro-mechanical technology are neither unique nor uniform across villages and cultivators. These relationships are further modified by various factors such as size of holding, draft power availability, etc.

Other considerations: Draft power availability and agro-climatic characteristics further condition the agro-mechanical technology diffusion. The latter mainly includes type of soil, variations in terrain, elevation, etc.

Low levels of prevailing technology and scientific knowledge at the local level limit production capabilities of the local technology generation system. These constraints are reflected in the lack of skill and fabrication facilities at the local level to produce and repair improved implements.[4]

All these factors in various combinations constrain or encourage adoption of specific agricultural implements. Conditions of production, existing agro-climatic conditions, fabrication and repair facilities and availability of draft power and complementary implements condition the process of agro-mechanical technology diffusion. In our study, inappropriateness of new implements with regard to these considerations emerged as the major factor constraining the diffusion of agro-mechanical technology and in providing impetus for various types of modifications in these implements.

Responses of technology users and generators at the local level

Considerable flexibility was observed in the combination of inputs used at the farm level. Potentially, every farmer may be somewhat different in the adaptation of techniques. We were able to capture some of these differences in our survey (chapter 4). The results of our survey clearly show that a dichotomous characterization of the diffusion process (adoption/non-adoption) may be misleading. Such a categorization obscures the nature and process of adaptation and fails to capture the extent and intensity of use of specific implements. Technology diffusion is a continuous process, and between adoption and non-adoption a range of options are available to the cultivators which exemplify the richness of the adaptation process at the farm level. Broadly, the process of agro-mechanical technology diffusion consists of the following changes (chapters 4 and 5):

1 Adoption of a new agricultural implement
2 Use of a modified version of a new agricultural implement
3 Use of a modified version of an old agricultural implement
4 Use of an old agricultural implement for a new purpose

The changes and modifications contained in these last three categories are aimed at reducing or eliminating operational disadvantages, widening the range of productive activities or processes for which the implement can be used, and emerged as possible solutions to the problems arising out of the various constraining factors enumerated earlier. It is important to note that for old and new implements which are broad substitutes for each other, if the last two options are feasible, the diffusion of new implements can be adversely affected. The adoption/adaptation/modification process takes place at two levels: at the level of users (cultivators) and at the level of fabricators (artisans). At his level, the cultivator primarily decides about 1 and 4 and the intensity of use of specific implements.[5] At the second level, the changes are induced by artisan-cultivator interaction and result in modifications of types 2 and 3. Some of these modifications are made by users as well. Changes of types 2 and 3 are tried in response to the various factors affecting technology diffusion and as solutions to operational problems with specific implements. The modifications typically involve the following types of changes:

a) Changes in the design
b) Removal of some parts
c) Addition of new parts
d) Substitution of parts/production material.

These changes are not mutually exclusive and, as discussed in chapters 4 and 5, combinations of such changes do occur. Also, these changes are specific to regions, cultivators and agricultural implements.

In most cases, solutions (in the form of changes noted in Tables 17 and 18 in chapters 4 and 5 respectively) to problems are *ad hoc* in nature and based on cultivators' perception of their needs and the available technological capability of the artisans in the locality. In many cases, modifications suggested by local artisans and peasants involved a combination of technological elements of traditional and improved agricultural implements. Modifications or adaptations documented during the survey reduced the operational disadvantages of both traditional and modern equipment, widened the range of productive activities for which implements (mainly traditional) could be used and improved the relative advantages of both traditional and new equipment over earlier versions. In other words, these modifications encouraged (or made possible) the persistence of some traditional implements and, at the same time, improved the adoptability of some modern ones. It is significant to note that cultivator-artisan interaction is the major source of adaptive improvements in agro-mechanical technology, Conspicuous by their absence are links between formal R&D and large-scale manufacturers with local artisans and cultivators. Similarly, state extension personnel do not contribute in a substantial way to the process of technology adaptation and diffusion appropriate to the changing needs of the region.

Conclusion

The small size of farm operations in backward regions and high man-land ratios limits the capacity to mechanize. In this context, improvements in smaller agricultural implements become relevant. Since the nature of agriculture differs across regions and the size of farm operations is very uneven, the market for agricultural implements is fragmented and not very large. Large-scale manufacturing of agricultural implements for such a market is not likely to be efficient as there will not be economies of scale in large-scale production.

The problems posed by small, scattered, fragmented and to an extent seasonal markets, are compounded because the distribution and servicing channels are not well organized. Besides, cultivator perceptions seem to differ from those of R&D personnel with respect to important problems. Likewise, forms in which cultivators prefer their needs to be satisfied also differ from region to region. Given the nature of agriculture, its diversity across regions and the role played by local artisans historically, the processes of adaptation, design modification, and even imitation or copying, must necessarily be at a decentralized level, in different agro-climatic regions without exchange of information or communication. Under prevailing market conditions, investments in large-scale fabrication of agricultural implements are likely to involve a waste of scarce capital.

For any decentralized fabrication activity to be a success, the needs of local users (cultivators, in this case) have to be ascertained. In market economies, the nature of users' demand is identified mainly by dealers and other commercial agents. Agents operating in a backward region do not seem to be able to play this role. The state machinery, represented by extension personnel, also does not make any attempt in this direction. Adaptation and modification taking place at the village level remain unassessed. It may be useful to analyse the *ad hoc* solutions of peasants and artisans at the local level and make an attempt to generalize the essential technological ingredients contained in these solutions.

In sum, various interacting factors, reflected in the conditions of production and agro-climatic characteristics of the region, do constrain the process of agro-mechanical technology diffusion. The cultivator-artisan interaction, nevertheless, provides scope for adaptive changes and innovation in agricultural implements. The strengthening of the artisan based fabrication networks and a combination of local experience and scientific knowledge through links with formal R&D can encourage the growth of local technological capabilities, speeding up the pace of technology diffusion.

Notes

1   For similar conclusions, see Binswanger and Shetty (1977).
2   See for example, the details of HYV maize and millet cultivation in Kalol villages (chapter 4).
3   See chapter 4 for cases which indicate the possibility of such factors at work. Elsewhere also it has been hypothesized that benefits for small farms from weed control activity may be low because the benefit of weed control under the existing crop varieties and fertilizer use are low. Alternatively, the fodder loss due to weed control, and the opportunity cost of spending time on weeding, on small farms is not compensated for by increases in yields (Binswanger and Shetty, 1977).
4   In some backward/tribal villages even good carpenters are not available and blacksmiths cannot serrate sickles (see chapter 4).
5   These decisions, however may be influenced by his interaction with others, e.g., fellow cultivators (chapters 4 and 5).

Appendix
Artisan Questionnaire

1. Identification of the Sample Household

(a) District:

(b) Taluka:

(c) Village:

(d) Serial No. of the Sample household:
     (According to the Houselist Schedule)

(e) Name of the head of the household:

(f) Name of the respondent:

(g) name of the investigator:

(h) Date of survey:

(i) Time interview began: Ended:

(j) More than one visit? Yes/No. If yes, reason?

(k) Name of the checker:

(l) Date of checking: Signature:

2. Household characteristics

(a) Caste:

(b) main source of household income during last year:

(c) Size of operational holding (hectares):

(d) Number of plots: Plot No: 1 2 3 4 5 6

(e) Area (in hectares):
     i)   Mortgaged in:
     ii)   Rented in:
     iii) Mortgaged out:
     iv)   Rented out

3. Details of Household Members

The following details about each member or the household were collected:

(a) Name

(b) Relation with the head of the household
(c) Age
(d) Sex
(e) Marital status (Code)
(f) Education (Code)
(g) main and subsidiary occupations (Activity, Industry and Occupation (see codes)

Code List (Block3)
Marital Status:
Unmarried (1); Currently married (2); widowed (3); separated (4);

Level of education:
Illiterate (1); Can read and write (2); Primary (3); Middle (4); Higher Secondary (5); Graduate (6); Technical Education, specify (6); Others, specify (7).

Activity Status:
Employer (1); Household enterprise (2); Helper in Household enterprise (3); Casual Labour (4); Regular wage earner or salaried employee (5); Did not work but was seeking employment (6); Domestic work (7); Student (8); Young, disabled, pensioners, beggars etc. (9).

Industry:
Agriculture and allied activities (0); Mining and quarrying (1); Household Industry (2); Other (non-household) industry (3); Electricity Gas and water (4); Construction (5); Wholesale and retail trade (6); Transport, storage and communications (7); Insurance, finance and business services (8); Community, Social and personal services (9).

Occupation:
Cultivator (0); Agricultural labour (1); Other labour (2); Fishermen, hunters, loggers and related workers (3); Professional, technical and related workers (4); Administrative executive and management workers (5); Clerical and related workers (6); Sales workers (7); Personal services (8); and Carpenters, blacksmiths, and other artisans (9).

4. Family History of the Household

The following information was collected about the grandfather, father and sons (not enumerated as members of the sample household) of the present head of the household:
(a) Main and secondary occupation
(b) Level of Education
(c) Whether stayed/worked in the same village?

(d) If no in (c), why did they move?
(e) Land owned
(f) If alive, present occupation
(g) Crops grown by them.

5. Implements Currently Fabricated

For each agricultural implement fabricated by the artisan, the following information was collected.
(a) Name of the implement/part
(b) Specification (Size, Wooden, iron etc.)
(c) Design source (code)
(d) Regularly produced or on order
(e) No. fabricated last year
(f) Current price (Rs)
(g) Raw Material
    i) Name
    ii) Quantity
    iii) Source
    iv) Cost
(h) Parts
    i) Name
    ii) Source
    iii) Price
(i) Since when are you making this implement/part?
(j) What category of cultivators buy this implement/part? (code)
(k) What type of repair or other service you provide for this implement?
(l) No. of days/hours taken to produce one unit
(m) Did your elders produce this implement/part?
(n) If yes, did they take the same time as you do? (code)
(o) If no, what are the reasons for this difference:
    i) Technical facilities?
    ii) Design difference?
    iii) Any other (specify)?
        (Please provide details).

Code List
Source of Design
Old ancestral design (1); Ancestral design modified by you (2); Design copied from some other artisan (3); Design copied from some implement available in the market (4); A new design of your own (5); Design given by Gujarat Agro-Industries Corporation (6); Any other, specify (7).

Category of cultivators
Marginal (1); Small (2); Medium (3); Large (4).

Time (code for 5 n)
Yes (1); No, less (2); No more (3); Do no know (4);
(For source of raw material, the artisan was specifically asked if the customer/cultivator provided it)

6. Fabrication Facilities

For each fabrication facility available to the artisan, the following details were collected.
(a) Name of the facility (machines, major tools, furnaces, grinders, lathes etc.)
(b) Number
(c) Power rating, if any (Horse Power)
(d) Source of energy:
    i) Power operated
    ii) Hand operated
    iii) Any other (specify)
(e) Year of purchase
(f) Cost (Rs.)
(g) Source of purchase
(h) Did your elders have this facility?
    i) Father   Yes/No
    ii) Grandfather   Yes/No

7. Loan/Subsidy

Along with the purpose of loan/subsidy the same information was collected as in Annexure I, Block 6.

8. Improvements in Agricultural Implements

Have you made any improvements in agricultural implements?
If yes, please give the following details:
(a) Name of the implement
(b) Improvement
    Specification (Details)
    i) Earlier
    ii) Now
(c) When was it introduced?
(d) Idea source for this improvement (code)
(e) How is it an improvement (code)?
(f) Has this change been accepted? Yes/No
    i) By artisans
    ii) By cultivators
(g) How did you popularise this improvement

Code List
Idea Source
Your experience (1); Cultivator/customer (2); Fellow artisan (3); Other
implements in the market (4); Any other, specify (5).

Improvement
Qualitatively/technically superior (1); Cheaper (2) due to: Cheaper material
(2a); new method/process of production (2b); Others, specify (2c).

9. Household Income and its sources:

Data on household income along with its cash/kind composition and
frequency of payment was collected separately for the following sources:
(a) Fabrication of agricultural implements
(b) Repair and maintenance of implements
(c) Crop income
(d) Dairy
(e) Wage income
    i) Agricultural
    ii) Others
(f) Trade
(g) Salary
(h) Others (interest, dividends, remittances, pension, allowances, rent,
income from trees, etc.)

10. General Questions

In addition to the structural questions enumerated above some semi-structured
and open-ended questions were also asked. We list them below.
(1) Since when are you making implements? Where did you receive your
training?
(2) In which villages you mainly supply your implements?
Please give names.
(3) In block 5 you have mentioned that you make some implements on the
basis of the specifications provided by the customer. What factors, in
your opinion, determine these specifications? Please provide details for
each implement separately.
(4) Do you meet the following persons? Yes/No
    (a) Fellow cultivators
    (b) Village artisan
    (c) Nearby village artisan
    (d) Nearby town artisan
    (e) Village level worker
    (f) Block development/extension officers
        If yes, do you discuss the following matters:
    (a) Problems with specific implements

(b) Possibilities of improving existing implements
(c) Designing new implements
(d) New implements which have recently come into the market
(e) A new design of an implement made by any artisan
     (Details were taken of any specific instance which came to the respondent's mind. If the response was No in most cases the respondent was asked to give his opinion as to why such interaction does not take place).

(5) When you get an idea about some improvement in an implement or about a new implement, what do you do?
     (If there has not been any instance of such type in the past, ask the artisan what he would do in case he gets an idea). Will you (Yes/No):
     (a) Make an experimental model and give it to cultivators for testing
     (b) Discuss the idea with:
          i)   Fellow artisans
          ii)   Cultivator friends and customers
          iii) Village level workers
          iv)   Taluka extension officers
          v)   Others (specify)
     (c) Anything else (specify)

(6) If the answer is negative to both questions (4) and (5), ask why in their opinion such discussions do not take place.

(7) If you do not make experimental models or pursue your ideas in some other fashion it is due to the lack of (Yes/No):
     (a) Resources
     (b) Fabrication and other facilities
     (c) Time
     (d) Interest on the part of the cultivators
     (e) Other (specify)
          (Please give details)

(8) What do you do if a new implement comes into the market and the demand for your implement declines?
     (If there are no instances of such type, ask what he will do if such a contingency arises). Will you (Yes/No):
     (a) Stop making that implement
     (b) Continue making it for a smaller clientele
     (c) Improve implement so that it can compete with the new one in the market
     (d   Start producing the new implement itself
     (e   Reduce the price of your implement.
          Cite instances, if any, of the type mentioned above.
          What problems you face/may face when you adopt the strategy (c) or (d).

(9) Do you take help from your family members in the production and sale of agricultural implements?

If yes,
- (a) How many of them help you?
- (b) During what season? (All, Kharif, Rabi, Summar)

(10) Do you hire labour? If yes, please give the following details for last year, separately for the three seasons if possible:
- (a) No. of hired workers
- (b) No. of days per worker
- (c) Wage rate (per worker per day/month)

(11) Do you undertake job work for outside firms? If yes, please give details of the goods you fabricate for them and their prices.

(12) Do you put out work? If yes, what work and to whom? Please give details.

(13) In the production of agricultural implements what kinds or problems do you face? Problems related with (give details):
- (a) Raw material
- (b) Fabrication facilities
- (c) Marketing
- (d) Credit facilities
- (e) Others (specify)

(14) What are your plans for the future? Do you plan to (Yes/No)
- (a) Fabricate more sophisticated agricultural implements
- (b) Other articles made of iron/wood
- (c) Others (specify)
     (Please give details)

(15) How do you fix the price of your implement?
    (The details of the process of price fixation, and differential prices for different category of customers were discussed with the respondent).

Bibliography

1. Basant, Rakesh (1985), 'Agricultural Technology and Employment in India: A survey of recent research', *Working Paper No.1*, The Gujarat Institute of Area Planning, Ahmedabad, India
2. Bath, B.H. Slicher Van (1960), 'The Influence of Economic Conditions on the Development of Agricultural Tools and Machines in History', in J.L. Meij (ed.) *Mechanisation in Agriculture*, North Holland Publishing Company, Amsterdam.
3. Bhalla, G.S. and Y.K. Alagh (1979), *Performance of Indian Agriculture*, A Districtwide Study, Sterling Publishers Pvt. Ltd.
4. Binswanger, H.P. and S.V.R. Shetty (1977), 'Economic Aspects of Weed Control in Semi-Arid Tropical Areas of India', *Occasional Paper No.13*, Economics Department, International Crops Research Institute for the Semi-Arid Tropics (ICRISAT) Hyderabad, India.
5. Binswanger, H.P. (1978), *The Economics of Tractors in South Asia — An Analytical Review*, Agricultural Development Council, New York and ICRISAT, Hyderabad, India
6. Binswanger, H.P. (1984), 'Agricultural Mechanization: A comparative historical perspective', *World Bank Staff Working Papers No.673*
7. Council for Advancement of Rural Technology (1985), *Rural Technologies Guide*, Publication 1, New Delhi, India
8. Dasgupta, Biplab (1977), *The New Agrarian Technology and India*, Macmillan
9. Feder, Gershon, Richard E. Just and David Zilberman (1982), 'Adoption of Agricultural Innovation in Developing Countries: A Survey', *World Bank Staff Working Papers No.542*
10. Government of Gujarat (1972), *Gujarat State Gazetteers, Panchmahals District*, Ahmedabad, India
11. Government of Gujarat (1975), *District Statistical Abstract, Panchmahals District, 1974-75*
12. Government of Gujarat, *Report of the Livestock Census, 1977* Bureau of

Economics and Statistics, Ghandinagar

13. Government of Gujarat (1979), *Report of the Gujarat State Land Commission*, Ghandinagar

14. Hale, S.M. (1973), *barriers to Free Choice in Development*, (Typed)

15. Hemmi, Kenzo (1982), 'Mechanization as a Strategy for Agricultural and Rural Development', Paper presented at the Seminar on 'Mechanization of Small Scale Farming held at Hangzhou, The Peoples Republic of China, June 22-26

16. India, national Commission on Agriculture, NCA (1976), *Report of the National Commission on Agriculture*, Ministry of Agriculture and Irrigation, Government of India

17. India, Planning Commission (1981), *Sixth Five Year Plan, 1980-85*, New Delhi

18. India, Registrar General 1971, Series 5, *Gujarat, District Census Handbook, Panchmahals District, Part X A and B Town and Village Directory*, Village and Townwise Primary Census Abstract

19. India, Registrar General (1982), *Census of India, 1981 Series-1 (India), Part II B (i), Primary Census Abstract, General Population*

20. International Labour Office (1974), *Mechaniztion and Employment in Agriculture*, Case Studies from Four Continents, Geneva

21. Joshi, P.C. (1979), 'Technological Potential of Peasant Agriculture', *Economic and Political Weekly*, Vol.XIV, No.26, June 30

22. Marsden, K. (1973), 'Technological Change in Agriculture, Employment and Overall Development Strategy' ILO (1973

23. National Sample Survey Organization, (NSSO), (1986), *Some Aspects of Household Ownership, Report on Land Holdings -1*, Thirty Seventh Round (1982), Department of Statistics, New Delhi, India

24. National Council of Applied Economic Research, (NCAER) (1981), *Implications of Tractorization for Farm Employment, Productivity and Income*, Summary and Highlights, NCAER, New Delhi, India

25. Raj, K.N. (1973), 'Mechanization of Agriculture in India and Sri Lanka (Ceylon) ILO (1973)

26. Sharan, Girija and P.V. Krishna (1976), *Problems in management, Custom Hiring Centres in Gujarat*, Indian Institute of Management Monograph, Ahmedabd, India

27. Sharma, U.S. (1983), 'Performance and Prospects of Gujarat Agriculture' in Lakdawala, D.T.(ed.,) *Gujarat Economy: Problems and Prospects*, Monograph Series 10, Sardar Patel Institute of Economic and Social Research, Ahmedabad, India

28. Vaidyanathan, A. (1982), *Role of Bovines in Indian Agriculture: A Study of Size Composition and Productivity*, Centre for Development Studies, Trivandrum, India

29. Vyas, V.S. (1976), 'Structural Changes in Agricuktural and Small Farm Sector' *Economic and Political Weekly*, Vol.XI, Nos. 1 and 2, January 19

30. Wang, I.K. (1981), 'Agricultural Technological Research and

Development in Developing Countries', *The Seoul National University Economic Review*, Vol.XV, Nol.1, December.